Problems and Solutions in Medical Physics

The third in a three-volume set exploring **Problems and Solutions in Medical Physics,** this volume explores common questions and their solutions in Radiotherapy. This invaluable study guide should be used in conjunction with other key textbooks in the field to provide additional learning opportunities. One hundred and forty-four solved problems are provided in ten chapters on basic physics topics, including External Beam Therapy Equipment, Photon Beam Physics, Radiation Dosimetry, Treatment Planning for External Beam Radiotherapy, and External Beam Commissioning and Quality Assurance. Each chapter provides examples, notes, and references for further reading to enhance understanding.

Features:

- Consolidates concepts and assists in the understanding and applications of theoretical concepts in medical physics
- Assists lecturers and instructors in setting assignments and tests
- Suitable as a revision tool for postgraduate students sitting medical physics, oncology, and radiological science examinations

Kwan Hoong Ng, PhD, DABMP, FIUPESM, is a Professor Emeritus, Faculty of Medicine, Universiti Malaya, Malaysia, and a Fellow of the IOMP. He received his MSc (Medical Physics) from the University of Aberdeen and PhD (Medical Physics) from the Universiti Malaya, Malaysia. He is certified by the American Board of Medical Physicists. Professor Ng was honoured as one of the top 50 medical physicists in the world by the International Organization of Medical Physics (IOMP) in 2013. He was the recipient of the Marie Sklodowska Curie Award in 2018. Professor Ng serves as a consultant for the International Atomic Energy Agency (IAEA).

Ngie Min Ung, PhD, MIPEM, MIPM, obtained his Bachelor of Biomedical Engineering and Master of Medical Physics from the Universiti Malaya, Malaysia. He then completed his PhD at the University of Western Australia. Dr. Ung is actively involved in medical physics teaching and research activities. To date, he has authored/co-authored over 70 peer-reviewed academic journal papers and supervised a total of 10 research PhD and master students to completion.

Robin Hill, PhD, MACPSEM, MAAPM, received his MSc (Medical and Health Physics) from the University of Adelaide and PhD from the University of Sydney. He is actively engaged in radiation oncology medical physics research, teaching and clinical activities including volunteer involvement in several IAEA activities. Dr. Hill has authored/co-authored more than 60 papers in peer-reviewed journals with a particular interest in therapeutic kilovoltage X-ray beam dosimetry. He is accredited in radiation oncology medical physics by the ACPSEM.

Series in Medical Physics and Biomedical Engineering

Series Editors:
Kwan Hoong Ng, E. Russell Ritenour, and Slavik Tabakov

For more information about this series, please visit: https://www.routledge.com/Series-in-Medical-Physics-and-Biomedical-Engineering/book-series/CHMEPHBIOENG

Problems and Solutions in Medical Physics

Radiotherapy Physics

Kwan Hoong Ng, Ngie Min Ung and
Robin Hill

CRC Press
Taylor & Francis Group
Boca Raton London New York

CRC Press is an imprint of the
Taylor & Francis Group, an **informa** business

First edition published 2023
by CRC Press
6000 Broken Sound Parkway NW, Suite 300, Boca Raton, FL 33487-2742

and by CRC Press
4 Park Square, Milton Park, Abingdon, Oxon, OX14 4RN

© 2023 Kwan Hoong Ng, Ngie Min Ung and Robin Hill

CRC Press is an imprint of Taylor & Francis Group, LLC

The right of Kwan Hoong Ng, Ngie Min Ung and Robin Hill to be identified as authors of this work has been asserted in accordance with sections 77 and 78 of the Copyright, Designs and Patents Act 1988.

This book contains information obtained from authentic and highly regarded sources. While all reasonable efforts have been made to publish reliable data and information, neither the author[s] nor the publisher can accept any legal responsibility or liability for any errors or omissions that may be made. The publishers wish to make clear that any views or opinions expressed in this book by individual editors, authors or contributors are personal to them and do not necessarily reflect the views/opinions of the publishers. The information or guidance contained in this book is intended for use by medical, scientific or health-care professionals and is provided strictly as a supplement to the medical or other professional's own judgement, their knowledge of the patient's medical history, relevant manufacturer's instructions and the appropriate best practice guidelines. Because of the rapid advances in medical science, any information or advice on dosages, procedures or diagnoses should be independently verified. The reader is strongly urged to consult the relevant national drug formulary and the drug companies' and device or material manufacturers' printed instructions, and their websites, before administering or utilizing any of the drugs, devices or materials mentioned in this book. This book does not indicate whether a particular treatment is appropriate or suitable for a particular individual. Ultimately it is the sole responsibility of the medical professional to make his or her own professional judgements, so as to advise and treat patients appropriately. The authors and publishers have also attempted to trace the copyright holders of all material reproduced in this publication and apologize to copyright holders if permission to publish in this form has not been obtained. If any copyright material has not been acknowledged please write and let us know so we may rectify in any future reprint.

ISBN: 9781482240054 (hbk)
ISBN: 9781032332789 (pbk)
ISBN: 9780429159466 (ebk)

DOI: 10.1201/9780429159466

Typeset in Minion Pro
by codeMantra

Contents

Preface

IN VIEW OF THE increasing number and popularity of master's and higher-level training programmes in medical physics worldwide, there is an increasing need for students to develop problem-solving skills in order to grasp the complex concepts which are part of the ongoing clinical and scientific practice. The purpose of this book is, therefore, to provide students with the opportunity to learn and develop these skills.

This book serves as a study guide and revision tool for postgraduate students sitting examinations in radiotherapy physics. The detailed problems and solutions included in the book cover a wide spectrum of topics, following the typical syllabi used by universities on these courses worldwide.

The problems serve to illustrate and augment the underlying theory and provide a reinforcement of basic principles to enhance learning and information retention. No book can claim to cover all topics exhaustively, but additional problems and solutions will be made available periodically on the publisher's website, www.routledge.com/9781032332789.

One hundred and forty-four solved problems are provided in ten chapters on basic physics topics, including External Beam Therapy Equipment, Photon Beam Physics, Radiation Dosimetry, Treatment Planning for External Beam Radiotherapy, and External Beam Commissioning and Quality Assurance.

The approach to the problems and solutions covers all six levels in the cognitive domain of Bloom's taxonomy.

This book is one of a three-volume set containing medical physics problems and solutions. The other two books in the set tackle diagnostic imaging physics and nuclear medicine physics.

We would like to thank the staff at Taylor & Francis, especially Rebecca Davies, Danny Kielty and Kirsten Barr, for their unfailing support.

Kwan Hoong Ng, Ngie Min Ung, and Robin Hill

Authors

Kwan Hoong Ng, PhD, DABMP, FIUPESM, is a Professor Emeritus, Faculty of Medicine, Universiti Malaya, Malaysia, and a Fellow of the IOMP. He received his MSc (Medical Physics) from the University of Aberdeen and PhD (Medical Physics) from the University of Malaya, Malaysia. He is certified by the American Board of Medical Physicists. Professor Ng was honoured as one of the top 50 medical physicists in the world by the International Organization of Medical Physics (IOMP) in 2013. He was the recipient of the Marie Sklodowska Curie Award in 2018. He has authored/co-authored over 280 papers in peer-reviewed journals, 30 book chapters and co-edited 12 books. He has presented over 500 scientific papers and more than 300 invited lectures. He has also organised and directed several workshops on radiology quality assurance, digital imaging and scientific writing. He has directed research initiatives in breast imaging, intervention radiology, radiological safety and radiation dosimetry. Professor Ng serves as a consultant for the International Atomic Energy Agency (IAEA) and previously served as a consulting expert for the International Commission on Non-Ionizing Radiation Protection (ICNIRP). He is the founding and emeritus president of the South East Asian Federation of Medical Physics (SEAFOMP) and is a past president of the Asia-Oceania Federation of Organizations for Medical Physics (AFOMP).

Ngie Min Ung, PhD, MIPEM, MIPM, is an associate professor and medical physicist at the Department of Clinical Oncology, University of Malaya. He obtained his Bachelor of Biomedical Engineering and Master of Medical Physics from the University of Malaya. He then completed his PhD at the University of Western Australia in collaboration with Genesis Cancer Care (formerly known as Perth Radiation Oncology). Dr. Ung is a fellow and committee member of Malaysia Association of Medical Physics and Institute of Physics Malaysia (Medical Physics Division). Besides providing medical physics services to his department, Dr. Ung is actively involved in medical physics teaching and research activities. To date, he has authored/co-authored over 70 peer-reviewed academic journal papers and supervised a total of ten research PhD and master students to completion. His PhD work revolved around the investigation of uncertainties in fiducial markers tracking during image-guided radiotherapy of prostate cancer. Apart from IGRT, his other current research interests include novel detectors for radiation dosimetry, brachytherapy and advanced radiotherapy techniques.

Robin Hill, PhD, MACPSEM, FAIP, grew up in Adelaide, Australia, where he completed his undergraduate studies and an MSc (Medical Physics) at the University of Adelaide. After graduation, he worked as a clinical radiation oncology medical physicist at a number of hospitals in Sydney, Australia. An interest in kilovoltage X-ray beam dosimetry that started during his honours project and clinical work leads to completion of his PhD from the University of Sydney. Subsequently, a review paper on kilovoltage X-ray beam dosimetry was published in Physics in Medicine and Biology in 2014 which has helped educate scientists around the world in this topic. Dr Hill was accredited in radiation oncology medical physics by the Australasian College of Physical Scientists and Engineers in Medicine. He has authored/co-authored more than 60 papers in peer-reviewed journals and presented at many conferences. Dr Hill is passionate about education in medical physics as evident in his involvement in teaching at the University of Sydney, including supervision of many students as part of their research projects. Dr Hill is also an examiner with the ACPSEM as part of the Radiation Oncology Examiners Panel. He is also involved in a number of radiation oncology projects within the IAEA supporting medical physics in South East Asia through on-site visits, PhD supervision and regional collaborations. When he is not working, Dr. Hill enjoys bike riding, travelling with his family and finding good coffee shops.

Acknowledgements

The authors would like to acknowledge the following contributors of this book:

Chai Hong Yeong, PhD, MIPM
School of Medicine
Faculty of Health and Medical Sciences
Taylor's University Lakeside Campus
Subang Jaya, Malaysia

Jeannie Hsiu Ding Wong, PhD, MIPM
Department of Biomedical Imaging
Faculty of Medicine
Universiti Malaya
Kuala Lumpur, Malaysia

Nur Diyana Afrina Mohd Hizam, MMedSc
Clinical Oncology Unit
Faculty of Medicine
Universiti Malaya
Kuala Lumpur, Malaysia

Tomas Kron, PhD, FCCPM, FCOMP, FACPSEM, FIOMP
Department of Oncology
Peter MacCallum Cancer Centre
Melbourne, Australia

Wei Loong Jong, PhD, DIMPCB, MACPSEM
Proton Therapy SG
Singapore Institute of Advanced Medicine
Singapore

Abbreviations

1D	One-dimensional
2D	Two-dimensional
3D	Three-dimensional
4DCT	Four-dimensional computed tomography
AAPM	American Association of Physicists in Medicine
ALARA	As low as reasonably achievable
BEV	Beam's eye view
CBCT	Cone-beam computed tomography
CI	Conformity index
CoP	Code of practice
CT	Computed tomography
CTDI	Computed tomography dose index
CTV	Clinical target volume
Co-60	Cobalt-60
DIBH	Deep inspirational breath hold
DICOM	Digital Imaging and Communications in Medicine
DNA	Deoxyribonucleic acid
DRR	Digitally reconstructed radiograph
DTA	Distance to agreement
DVH	Dose volume histogram
EPID	Electronic portal imaging device
EQD2	Equivalent dose in 2-Gy fractions
FFF	Flattening filter free
FOV	Field of view

FSU	Functional subunits
GM	Geiger-Muller
GTV	Gross tumour volume
HDR	High-dose rate
HVL	Half-value layer
I-131	Iodine-131
IAEA	International Atomic Energy Agency
ICRP	International Commission on Radiological Protection
ICRU	The International Commission on Radiation Units and Measurements
IEC	International Electrotechnical Commission
IGRT	Image-guided radiotherapy
IMRT	Intensity-modulated radiotherapy
Ir-192	Iridium-192
ITV	Internal target volume
IV	Irradiated volume
keV	Kiloelectron volt
kV	Kilovoltage
kVCT	Kilovoltage computed tomography
LCPE	Lateral charged particle equilibrium
LDR	Low-dose rate
LET	Linear energy transfer
LMA	Low-melting alloy
LNT	Linear no-threshold
LQ	Linear quadratic
MeV	Megaelectron volt
MRI	Magnetic resonance imaging
MV	Megavoltage
MLC	Multi-leaf collimator
MOSFET	Metal–oxide–semiconductor field-effect transistor
MU	Monitor units
MVCT	Megavoltage computed tomography
NTCP	Normal tissue control probability

OAR	Organ at risk
OBI	On-board imager
OER	Oxygen enhancement ratio
OF	Output factor
Os-192	Osmium-192
OSLD	Optically stimulated luminescence dosimeter
P-32	Phosphorus-32
PDD	Percentage depth dose
Pt-192	Platinium-192
PMMA	Polymethyl methacrylate
PSDL	Primary standard dosimetry laboratory
PTV	Planning target volume
QA	Quality assurance
QUANTEC	Quantitative analysis of normal tissue effects in the clinic
R & V	Record and verify
RBE	Relative biological effectiveness
RF	Radiofrequency
RMP	Radiation management plan
ROSIS	The Radiation Oncology Safety Information System
SABR	Stereotactic ablative radiotherapy
SAD	Source-to-axis distance
SAFRON	Safety in Radiation Oncology
SBRT	Stereotactic body radiotherapy
SGRT	Surface-guided radiotherapy
SLD	Sublethal damage
SRS	Stereotactic radiosurgery
SSD	Source-to-surface distance
SSDL	Secondary standard dosimetry laboratory
TBI	Total body irradiation
TCP	Tumour control probability
TLD	Thermoluminescent dosimeter
TMR	Tissue maximum ratio

TPR	Tissue phantom ratio
TPS	Treatment planning system
TV	Treated volume
TVL	Tenth value layer
VMAT	Volumetric arc therapy
WFF	With flattening filter

CHAPTER 1

External Beam Therapy Equipment

1.1 X-RAY MACHINE

Problem:
Treatment of cancers using X-rays is an important modality. Give the range of electron kinetic energies for the following X-ray modalities:

- a. Superficial X-ray machine
- b. Orthovoltage machine
- c. Megavoltage machine

Solution:

- a. 10–100 keV
- b. 101–500 keV
- c. > 1 MeV

DOI: 10.1201/9780429159466-1

1.2 Co-60 UNIT

Problem:
Linear accelerators are the most common megavoltage units for external beam radiotherapy treatments. They have been developed to replace Co-60 and kilovoltage treatment machines.

a. List the advantages of linear accelerators over Co-60 machine.

b. What is the purpose of a 'trimmer' in the radiotherapy treatment using Co-60 machine?

Solution:

a. Advantages:

- Linear accelerator can provide either megavoltage electron or X-ray therapy with a wide range of energies, allowing radiation oncologists to tailor treatment to the required depth.

- Most modern linear accelerator features such as high dose rate modes, MLC, electron arcs therapy, dynamic wedges and dynamic MLC operation during treatment.

- No radioactive source contamination.

- Higher dose rate (1–10 Gy min^{-1}) allowing shorter treatment time.

- Have a sharp dose fall-off at the beam edge than Co-60 beam.

b. The trimmer consists of heavy metal bars used to attenuate the beam in the penumbra region, thus 'sharpening' the field edges.

1.3 LINEAR ACCELERATOR

Problem:
Draw a 2D sketch and name the components of a single energy medical linear accelerator from the electron gun to the collimator jaws.

Solution:
See Figure 1.1.

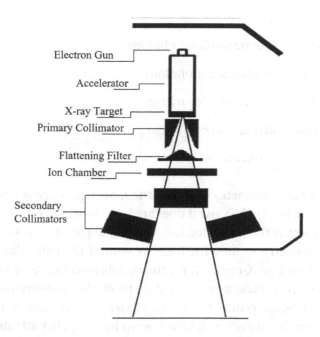

Electron Gun

Accelerator

X-ray Target
Primary Collimator

Flattening Filter
Ion Chamber

Secondary Collimators

FIGURE 1.1 Schematic diagram of a linear accelerator head.

1.4 NARROW VERSUS BROAD BEAM GEOMETRY

Problem:
Explain the difference between narrow beam and broad beam geometry for photon beams? Give examples of their use in radiotherapy.

Solution:
Narrow beam geometry involves dose measurements for a narrow (or small) beam where you are only measuring primary photons from the incident beam. In the ideal situation, this means that there is no scattered radiation in the beam from the attenuator. From this, the linear attenuation coefficient (μ) can be determined based on the equation:

$$N = N_0 \times e^{-\mu x}$$

where

N = number of transmitted photons

N_0 = number of incident photons

e = the base of natural logarithm

μ = linear attenuation coefficient

x = absorber thickness

Narrow beam geometry produces the most reproducible attenuation coefficient (μ) because it does not include field size and scatter effects; both of them are effectively zero. Examples of uses of narrow beam geometry include the measurement of the half-value layer on a kilovoltage X-ray unit, for the measurement of mass energy attenuation coefficients and radiation protection measurements.

Broad beam geometry allows scatter and secondary radiation to reach a detector. A broad beam has a smaller attenuation coefficient than a narrow beam. Some of the attenuation is 'cancelled out' by scatter; therefore, the coefficient is smaller. A broad beam has a thicker HVL than a narrow beam. A thicker barrier is required to shield against a broad beam as both primary beam and scatter must be shielded (as opposed to just the primary beam). Examples of uses of broad beam geometry include radiation shielding calculations for linear accelerator bunkers and radiation protection measurements such as dose surveys.

1.5 PHOTON AND ELECTRON TREATMENT MODES

Problem:

What differences occur in and on the linear accelerator treatment head when X-ray mode is deactivated and electron therapy mode is activated?

Solution:

The target and flattening filter are removed from the beam line and a scattering foil is placed in the electron path (Figure 1.2). Instead of striking the target, the electrons will strike the scattering foil resulting in beam spreading and uniform electron fluence across the field. The scattering foils are usually made up of two thin metallic foils and are different for different beam energies. They are sufficiently thin to enable most electrons to be scattered instead of generating bremsstrahlung. However, a small amount of bremsstrahlung occurs and appears as photon contamination of the electron beam which increases with electron energy. The emitting electrons are charged particles and tend to interact more with air than photons, and this leads to energy degradation of the somewhat monoenergetic electron beams. The angular scattering of the electron beams also increases as the electrons traverse away from the scattering foil.

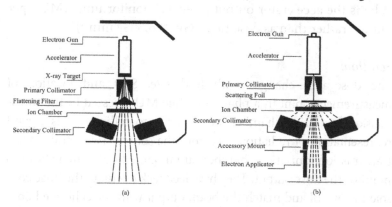

FIGURE 1.2   Schematic drawing of (i) target and flattening filter and (ii) scattering foil in the beam line.

Hence, the use of electron applicators is necessary to collimate electron beams as near as possible to the patient.

1.6 ANGULAR DISTRIBUTION AND EFFICIENCY OF X-RAY POPULATION

Problem:
Compare the X-ray production for a kilovoltage and a megavoltage X-ray unit in terms of angular distribution of photons and efficiency of X-ray production.

Solution:
High-energy X-rays are mainly forward directed, while low-energy X-rays are primarily emitted perpendicular to the incident electron beam. The higher the energy, the more efficient the X-ray production. At low kilovoltage energies (~100 kV), the efficiency of X-ray production is approximately 1% where most of the energy of the electrons (about 99%) is converted into heat. At megavoltage energies (~20 MV), the efficiency of X-ray production is approximately 70%.

1.7 MONITOR UNITS

Problem:
Why is the accelerator output given as 'monitor units (MU)' per minute rather than as dose rate (cGy s^{-1} or cGy min^{-1})?

Solution:
The dose rate changes with field size, SSD and location of measurement point in a phantom while MU assessed in the transmission ionisation chamber located prior to collimation is a good representation of radiation flux from the source. Two parallel plate transmission ionisation chambers are mounted in the linac head to monitor the dose output. They have feedback circuit to the linac console to control and match the beam output with the calibrated dose rate. The ionisation chambers are called monitor units ionisation chambers; hence, the output is called monitor units. MU represents

the dose delivered to the patients, as long as they are calibrated. The most common calibration setup is that 1 MU is usually calibrated to be equal to 1 cGy of absorbed dose delivered at 100 cm SSD for a 10 cm × 10 cm field size at d_{max} in water. However, this may vary according to the equipment and setup in the local radiotherapy department.

1.8 RADIATION YIELD

Problem:
Define 'radiation yield' in the production of photon beams in a linear accelerator.

Solution:
Radiation yield refers to the total fraction of the energy that is emitted as electromagnetic radiation (bremsstrahlung), while the electrons decelerate and come to rest.

In a linear accelerator, the efficiency of this process is dependent on the energy of the electrons and the target materials used. Not all electrons are converted to X-ray photons, so the rest of the electrons lose energy through collision with the atoms of the target material through ionisation and excitation. Energy is deposited in the material resulting in the intense heat of the target material. Therefore, an efficient cooling system is needed to remove the heat at the target.

1.9 TARGET

Problem:
What material properties should a 6 MV linear accelerator target have? Give an example of a suitable target material.

Solution:
Suitable materials for linac targets should consist of materials of high atomic number (Z), high melting point and good thermal conductivity. A good example is tungsten (W) which has relatively high Z ($Z = 74$), melting point of 3422°C and thermal conductivity of 173 W·m⁻¹ K⁻¹. The higher density of tungsten (19.3 g.cm⁻³) also helps

as it confines the radiation source to a smaller physical location. Lead (Pb) is not a suitable target material because although it has very high Z (82), it has low melting point and poor thermal conductivity (Pb = 327.46°C, 35.3 W·m^{-1} K^{-1}).

1.10 LINEAR ACCELERATOR DOSE RATE

Problem:
The radiation dose delivered by any therapeutic machine is often described by the average dose rate in the units of Gy min^{-1}. Contrast this to a Co-60 unit, linear accelerator delivers radiation in bursts of radiation beam pulses; hence its dose rate can be specified in dose per radiation beam pulse, sometimes also referred to as the instantaneous dose rate.

A linear accelerator is delivering at a dose rate of 4 Gy min^{-1} to a calibration point in water. The radiation beam is delivered in 3.5 μs pulses and the pulse period is 2.8 ms. See the radiation beam pulse diagram in Figure 1.3.

  a. What is the average dose rate of the beam?

  b. What is the instantaneous dose rate or the dose per radiation beam pulse of the beam?

Solution:
  a. Average dose rate = 4 Gy min^{-1}.

  b. Dose per second, $4\,\mathrm{Gy\,min^{-1}} = \dfrac{4\,\mathrm{Gy}}{60\mathrm{s}} = 0.067\,\mathrm{Gy\,s^{-1}}$

$$\text{No. of pulse per second} = \frac{1\mathrm{s}}{2.8\mathrm{ms}} = 357 \text{ pulses}$$

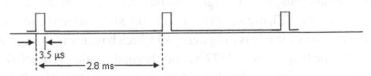

FIGURE 1.3   Radiation beam pulse diagram.

Therefore, the instantaneous dose rate is

$$\text{Dose per radiation pulse} = \frac{0.067 \text{ Gy}}{357 \text{ pulses}} = 1.867 \times 10^{-4} \text{ Gy pulse}^{-1}$$

1 dose pulse = 1.867×10^{-4} Gy is delivered in 3.5 μs

$$\text{Dose/second} = \frac{1.867 \times 10^{-4} \text{ Gy}}{3.5 \text{ μs}} = 53.34 \text{ Gy s}^{-1}$$

Dose/min = 53.34 Gy s^{-1} × 60s = 3200 Gy min^{-1}

1.11 MAGNETRON VERSUS KLYSTRON

Problem:
Compare and contrast the use of a magnetron and a klystron in a radiotherapy linear accelerator (Figures 1.4 and 1.5).

Solution:
Magnetron and klystron are both devices that produce/amplify microwave signal in a medical linear accelerator. The schematic drawings of a magnetron and klystron are shown in Figure 1.4

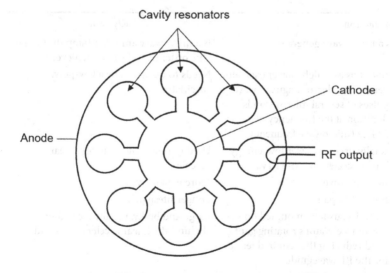

FIGURE 1.4 Cross-sectional schematic drawing of a magnetron.

and Figure 1.5 respectively. The differences between a magnetron and a klystron are summarised in the table below.

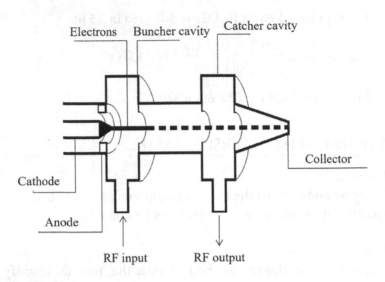

FIGURE 1.5   Schematic drawing of a klystron.

Magnetron	Klystron
Is a microwave generator	Is a microwave amplifier (amplification in the order of 10^{11} can be achieved)
Functions as a high-power oscillator generating electromagnetic wave pulses of several microseconds' duration at the frequency of ~ 3 GHz (microwave frequency)	Needs to be driven by a low-power oscillator
Usually used to power low-energy linear accelerators (≤ 6 MV)	Used to power high-energy linear accelerators
Less expensive	More expensive
Shorter lifespan	Longer lifespan
Small size, usually mounted on the linear accelerator's rotating gantry head reducing the required length for the RF waveguide	Large and bulky, usually located in or behind the linear accelerator's stand
Peak power ~2 MW	Peak power ~7 MW

1.12 KLYSTRON

Problem:
Describe briefly the operation of a klystron in a medical linear accelerator.

Solution:
Unlike magnetron, a klystron is not a source of microwaves. It is rather an amplifier of microwaves that is driven by source of low-energy oscillators. The electron gun produces a flow of electrons from the cathode. These electrons are accelerated by a negative pulse of voltage into the first cavity. This cavity is called the buncher cavity, which is energised by low-power microwaves. The microwaves generate an alternating electric field across the cavity. The velocity of the electrons is modified by a process known as the velocity modulation through the action of this electric field. This results in bunching of electrons as the velocity-modulated beam passes through a field-free space in the drift tube.

The electrons that pass through when the electric field opposes their motion are slowed, while electrons which pass through when the electric field is in the same direction are accelerated. As the electron bunches arrive at the catcher cavity, they induce charges on the ends of the cavity and thereby generate a retarding electric field. The electrons suffer deceleration, and by the principle of conservation of energy, the kinetic energy of electrons is converted into high-power microwaves. The spent electron beam, with reduced energy, is captured by a collector electrode. To make an oscillator, the output cavity can be coupled to the input cavity(s) with a coaxial cable or waveguide. Positive feedback excites spontaneous oscillations at the resonant frequency of the cavities. Most high-energy linacs (>6 MV) use klystrons.

1.13 WAVEGUIDE

Problem:
What is a waveguide? Name two types of accelerator waveguide designs commonly used in a medical linear accelerator and state their main differences.

Solution:

Waveguides are evacuated or gas-filled metallic structures of rectangular or circular cross-sections used in transmission of microwaves. The simplest kind of an accelerating waveguide is obtained from a cylindrical uniform waveguide by adding a series of disks (irises) with circular holes at the centre, placed at equal distances along the tube. These disks divide the waveguide into a series of cylindrical cavities that form the basic structure of the accelerating waveguide in a linac. The cavities serve two purposes, i.e., (i) to couple and distribute microwave power between adjacent cavities and (ii) to provide a suitable electric field pattern for acceleration of electrons. Two types of accelerating waveguides have been developed for acceleration of electrons, namely, the travelling waveguide and the standing waveguide.

The main differences between the travelling waveguide and the standing waveguide are listed in the table below.

Travelling Waveguide	Standing Waveguide
Travelling waveguides are generally longer physically.	The waveguide design may include side cavities which allow for the manufacturing of a shorter waveguide while still maintaining excellent acceleration.
At the end of the waveguide microwaves are absorbed without any reflection and fed back to the input.	At the end of the waveguide microwaves are reflected back to the input.
It requires low microwave peaked power (e.g., 2 MW).	It requires higher microwave peaked power than the travelling waveguide (e.g., 2.5 MW).
It requires lower mean RF power.	It requires higher mean RF power (~25% more).
It is used in Elekta linear accelerators.	It is used in Varian linear accelerators.

1.14 LINEAR ACCELERATOR BEAM ENERGY

Problem:
Explain briefly what factors affect the linear accelerator beam energy.

Solution:

- RF power – As the RF input power increases, the X-ray beam energy increases.

- Gun current/beam loading – As the gun current increases, the number of electrons released into the accelerating waveguide increases. This results in a higher current. If the power is identical, higher current results in lower voltage and thus energy; $P = I \times V$.

- Beam magnet shunt values – In linear accelerators with a bending magnet, changes in shunt voltages will affect the beam energies that pass through the 3% energy window. The 3% energy slits only allow electrons with the proper energies to pass through. The electrons outside this window strike the energy slits, removed from the pencil beam of electrons and are converted to heat.

- Length of waveguide – The longer the waveguide, the higher energy the electrons can be accelerated as the number of cavities in which acceleration takes place increases. The cavity size is determined by the wavelength of the RF, and it is a fixed value. This principle is used by introducing an 'energy switch' that effectively alters the length of the accelerating wave guide.

1.15 ENERGY SWITCH

Problem:
Dual energy linear accelerators with standing wave guides often use an energy switch to change the beam energies. Describe briefly

the operating principle of an energy switch in the accelerator waveguide.

Solution:
The energy switch introduces a phase shift for the RF wave by mechanically being inserted into a side coupling cavity. The phase of the RF power for every cavity after the energy switch is now out of alignment modifying the amount of energy electrons can take up in the cavity. No electron acceleration after the energy switch is possible. The accelerating electric fields in the bunching cavities are unchanged, and electrons are accelerated with little energy spread.

1.16 MULTILEAF COLLIMATORS

Problem:
Discuss the advantages of using multileaf collimators (MLC) over shielding blocks. Give some examples where the use of shielding blocks could still provide advantages for patient treatment.

Solution:
Advantages of using MLC:

- Automatic positioning of MLC to the required shape reduces the time for setup and makes multiple shaped fields more viable.

- MLC fields allow for delivery of radiotherapy techniques such as IMRT and VMAT.

- Using MLC reduces the time and hazard of lifting heavy blocks and mounting them in the collimator head above the patient.

- Significant block preparation time spent shaping moulds, then pouring low melting point allow (LMA) and mounting the blocks on block trays is now replaced with MLC.

- Time spent attaching blocks to block trays are avoided with MLC.

However, there are still some situations where blocks are used instead of MLC. These include:

- Island blocking

- Spinal cord shielding to avoid MLC end leaf leakage

- Secondary blocks to further reduce leakage for highly sensitive structures (e.g., embryo in pregnant patients undergoing radiotherapy)

- Electron applicators require LMA inserts for patient treatments

CHAPTER 2

Photon Beam Physics

2.1 INTERACTIONS OF X-RAY WITH MATTER (1)

Problem:
Describe briefly the following photon interaction mechanisms:

a. Compton effect

b. Photoelectric effect

c. Pair production

For each interaction above, list two factors that influence the probability of an interaction occurring.

Solution:

a. The Compton effect (incoherent scattering) represents a photon interaction with an essentially 'free and stationary' orbital electron. The incident photon energy is much larger than the binding energy of the orbital electron. The incident photon loses part of its energy to the recoil (Compton) electron and is scattered as photon with lower energy through a scattering angle. Two factors that influence the probability

DOI: 10.1201/9780429159466-2

of Compton effect are the energy of incident photon and electron density of interacting medium.

b. In the photoelectric effect the photon interacts with a tightly bound orbital electron of an attenuator and is completely absorbed, while the orbital electron is ejected from the atom as a photoelectron. Two factors that influence the probability of photoelectric effect are the energy of incident photon and the atomic number of interacting medium.

c. When the energy of an incident photon is greater than the rest mass energy of an electron and a positron ($2m_0c^2 = 1.022$ MeV), the photon may be absorbed through the mechanism of pair production. Here, the photon passes very close to the nucleus, and its energy is used to create an electron-positron pair. The excess energy is shared between the electron and positron as kinetic energy. These two particles may collide through a process known as annihilation and converted to two photons with equal energy of 511 keV. Two factors that influence the probability of pair production are the energy of incident photon and the atomic number of interacting medium.

2.2 INTERACTIONS OF X-RAY WITH MATTER (2)

Problem:
Explain briefly how Rayleigh scattering, the photoelectric effect, Compton scattering and pair production may vary as a function of photon energy and atomic number of the medium.

Solution:
Rayleigh scattering mainly occurs at low photon energies and the interaction cross-section data:

- Decreases very quickly as the photon energy increases and

- Increases with increasing atomic number Z according to a function called the atomic form factor.

Photoelectric effect interaction cross-section data:

- Varies with the photon energy, E, approximately as $1/E^3$.
- Varies with the atomic number of the medium by approximately Z^3 between to Z^4.

Compton scattering:

- Varies and slowly decreases with an increase in photon energy.
- Is almost independent of the atomic number Z.

Pair production:

- Occurs at minimum photon energy of 1.022 MeV.
- Increases quickly with increasing photon energy greater than 1.022 MeV.
- Varies with the atomic number of the medium by approximately Z^2.

2.3 INTERACTIONS OF X-RAY WITH MATTER (3)

Problem:
Figure 2.1 shows the mass energy absorption coefficients for air, cortical bone, lead and water as a function of X-ray energy ranging from 1 keV to 20 MeV. Explain the main reasons for the differences in the coefficients.

Solution:
It is noted that these are the mass energy absorption coefficients $\left(\dfrac{\mu}{\rho}\right)_{en}$ and as such the effects of density have been considered in the calculation. At the lower energies, the photoelectric effect is the dominant interaction process which has a strong dependence

X-ray mass energy absorption coefficients

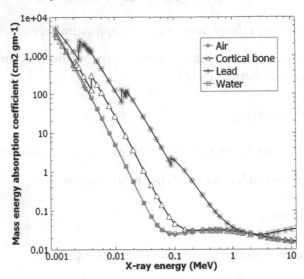

FIGURE 2.1 The X-ray mass energy absorption coefficients for air, cortical bone, lead and water as a function of X-ray energy ranging from 1 keV to 20 MeV (image taken from https://www.nist.gov/pml/X-ray-mass-attenuation-coefficients).

on the atomic number Z and inversely proportional to the X-ray energy. This explains why the value of $\left(\dfrac{\mu}{\rho}\right)_{en}$ is greatest for the lead, then cortical bone, water and air.

For the lead, one can see the K-, L- and M-edges which occur at the binding energies of the inner most electrons. The various edges are also readily visible in this plot for the cortical bone.

As the X-ray energy increases, the Compton effect becomes the dominant interaction process. The Compton effect has a much smaller variation with X-ray energy from around 200 keV to 10 MeV for both the water and cortical bone. This is the dominant interaction process in conventional radiotherapy using megavoltage X-ray beams or Co-60.

As the X-ray energy increases, pair production starts to become more important due to its dependence on both the X-ray energy E and the atomic number of the material. This explains the increase in the value of $\left(\dfrac{\mu}{\rho}\right)_{en}$ for lead from around 4 MeV.

2.4 BREMSSTRAHLUNG X-RAY

Problem:

A high proportion of the X-ray produced in a linear accelerator and therapeutic kilovoltage X-ray unit consists of bremsstrahlung photons.

a. Describe briefly the principle of bremsstrahlung X-ray generation.

b. Give three factors affecting the production of a bremsstrahlung spectrum.

c. Draw and label a typical kV X-ray spectra for:

   i. Unfiltered spectrum (spectrum out of the X-ray anode)

   ii. Spectrum with inherent filtration

   iii. Spectrum with total filtration (additional filtration)

Solution:

a. Bremsstrahlung X-ray resulted from Coulomb interactions between the incident electron and the nuclei of the target material. During the Coulomb interaction between the incident electron and the nucleus, the incident electron is decelerated (Bremsen is German for 'braking') and loses part of its kinetic energy in the form of bremsstrahlung photons.

b. Kinetic energy of electrons, thickness of target and atomic number of target.

c. See Figure 2.2.

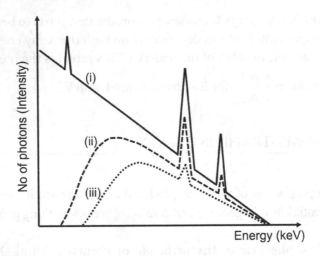

FIGURE 2.2   Typical X-ray spectra of (i) unfiltered spectrum, (ii) spectrum with inherent filtration and (iii) spectrum with total filtration.

2.5 INVERSE SQUARE LAW

Problem:
A linear accelerator generates a 6 MV X-ray beam with a nominal dose rate in air of 6 Gy per minute at the isocentre position, i.e., SSD = 100 cm. Calculate the expected dose rate in air at distances of 80, 120 and 300 cm distances. Note any assumptions made in your calculations.

Solution:
In this case, we use the inverse square law equation where the dose rate drops inversely as the square of the distance. We assume that there is no attenuation and scattering of the X-ray beam in air but note that strictly this is only strictly true in the case of a vacuum.

The dose rate at a different SSD can be calculated by the following formula:

$$\frac{\text{Dose rate at new SSD (cm)}}{\text{Dose rate at 100 cm}} = \frac{\left(\text{SSD} = 100 \text{ cm}\right)^2}{\left(\text{New SSD in cm}\right)^2}$$

Therefore,

$$\text{Dose rate at new SSD (cm)} = \text{Dose rate at 100 cm} \times \frac{100^2}{(\text{New SSD in cm})^2}$$

Solving for SSDs of 80, 120 and 300 cm by using the above formula, the expected dose rates in air are given in the following table:

SSD (cm)	Dose Rate in Air (Gy per minute)
80	9.375
120	4.167
300	0.667

2.6 DEFINITION OF PERCENTAGE DEPTH DOSE, TISSUE PHANTOM RATIO AND TISSUE MAXIMUM RATIO

Problem:

With the aid of a formula and a schematic diagram showing the relative beam source position and patient/phantom surface define the following terms:

a. Tissue phantom ratio (TPR)

b. Tissue maximum ratio (TMR)

c. Percentage depth dose (PDD)

Solution:

a. Tissue phantom ratio (TPR) (Figure 2.3)

- The TPR is defined as the ratio of absorbed dose at any given point (d_{ref_2}) to the absorbed dose at the same distance from the source but with the surface of the phantom moved so that this point is at a specified reference depth (d_{ref_1}).

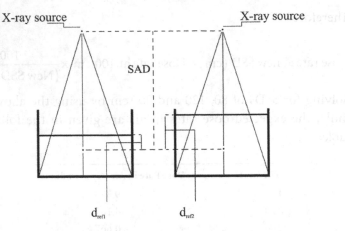

FIGURE 2.3   Setup for TPR measurement.

- Typical reference depth would be 10 cm which is also the depth recommended in most protocols for beam calibration.

b. Tissue maximum ratio (TMR) (Figure 2.4)

- Special case of TPR where d_{ref} is at the depth of maximum dose (d_{max}) (Figure 2.4).

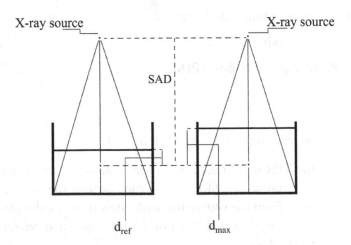

FIGURE 2.4   Setup for TMR measurement.

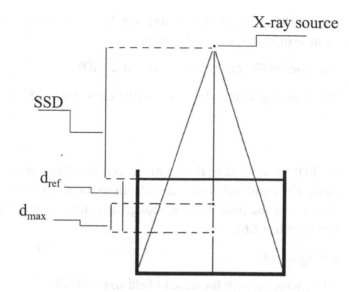

FIGURE 2.5   Setup for PDD measurement.

$$TMR\,(d_2)= \frac{D_{tissue}\,(d_{ref2})}{D_{tissue}\,(d_{max})}$$

c. Percentage depth dose (PDD) (Figure 2.5)

- PDD is the ratio of absorbed dose at any point to the dose at d_{max} within a water phantom at a fixed SSD (usually 100 cm) (Figure 2.5).

- $\%D\big(d_{ref},f,S,E\big)=\dfrac{D\big(d_{ref},f,S,E\big)}{D\big(d_{max},f,S,E\big)}\times100\%$

2.7 PERCENTAGE DEPTH DOSE

Problem:
PDD is an important measurement to characterise dose distribution.

a. Define PDD

b. Sketch the percentage depth dose distributions of an ortho-voltage X-ray unit, a Co-60 unit and a linac generating X-ray beam with energies of 4 MV and 10 MV. Assume that doses are measured in a water phantom.

c. How does PDD vary with field size and SSD?

d. How does the surface dose vary with field size and SSD?

Solution:

a. The PDD is defined as the quotient, expressed as a percentage, of the absorbed dose at any depth to the absorbed dose at depth of maximum dose, along the central axis of the beam (Figure 2.6).

b. See Figure 2.6.

c. PDD increases with increase in field size and SSD.

d. Surface dose increases with field size and reduces with SSD (more noticeable at larger field sizes due to reduced electron contamination from the treatment head).

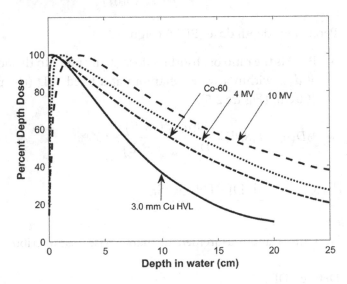

FIGURE 2.6 Comparative central-axis dose distributions of an orthovoltage X-ray unit, a Co-60 unit and a linac generating X-ray beam with energies of 4 and 10 MV.

2.8 HORNS OF A PROFILE

Problem:

With the help of a diagram, describe X-ray beam horns in a 6 MV dose profile. Explain why X-ray beam horns are evident at d_{max} but disappear at 10 cm depth.

Solution:

The profiles of 6 MV X-ray at d_{max} and at 10 cm depth are shown in Figure 2.7. X-ray beam horns exist due to the use of a flattening filter to produce a homogenous dose profile across the whole useable beam (typically $40 \times 40 \, cm^2$) at 100 cm distance from the target with the forward-peaked photon emission from the target. The flattening filter is a conical-shaped metal filter made from a high atomic number material. When the photon beam passes through the centre of the flattening filter, the photons get attenuated more compared to those that passes through the thin edge of the flattening filter. This produces the dose horns at shallower

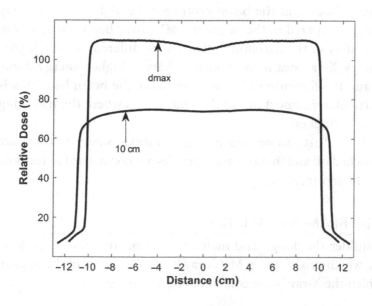

FIGURE 2.7 Profiles of 6 MV X-ray at d_{max} and at 10 cm depth.

depths such as d_{max}. The flattening filter is designed to produce a flat field at a specific depth, usually at 10 cm depth in water in line with IEC guidelines. Hence, at other depths, the dose profile may have horns or sloping shoulders. The X-rays that traverse the thinner part of the flattening filter have more low energy components. At greater depths, the softer X-rays penetrating the outer circle of the flattening filter are preferentially absorbed by water or tissue.

2.9 BEAM PROFILE

Consider a linear accelerator which can produce a 6 MV X-ray beam. The incident electrons that are directed towards the target have energy of 6 MeV. Explain what would happen to the beam profile if the energy of the electrons were to increase to 10 MeV.

Solution:
The higher energy electrons will produce a more forward-directed X-ray beam. This means that when profiles are measured in a scanning water tank, the horns in those profiles will be reduced. The detector used in the water tank will measure a larger dose on at the beam centre for the higher energy X-ray beam compared to the original 6 MV X-ray beam. In addition, the effect of the flattening filter will be different for the higher energy X-ray beam. As you now have a higher energy X-ray beam, the flattening filter will attenuate the beam less. This is particularly important on the central axis where the flattening filter is thickest.

Note: This answer assumes that various safety features are switched off and that the beam profiles are normalised to the dose on the central axis.

2.10 BEAM ASYMMETRY

Consider the design and main components of a linear accelerator which produces a 6 MV X-ray beam. List some of the ways in which the X-ray beam could become asymmetrical.

Solution:
Some of the possible reasons include the following:

- The X-ray target is not correctly aligned or centred on the central axis.

- The incident electron beam on target is not correctly aligned in terms of its position or angle of incidence.

- There are issues with the waveguide and bending magnet system (note some of this may result in the previous answer).

- The monitoring ionisation chamber in the gantry head may have issues with some of the lateral or transverse chamber segments.

2.11 FLATTENING FILTER-FREE BEAM PROFILE

Problem:
Flattening filter-free (FFF) photon beams are becoming much more widely available on radiotherapy linear accelerators. Figure 2.8 shows depth doses and lateral profiles for 10 MV photon beams: one with flattening filter (WFF) and one FFF.

a. What are the main dosimetric properties of FFF X-ray beams?

b. Explain the reason for the difference in the percentage depth doses and profiles.

c. List possible applications for FFF X-ray beams.

Solution:

a. The two main dosimetric properties of FFF beams are:

- A cone shapes profile with the maximum dose on the central axis and then drops off laterally

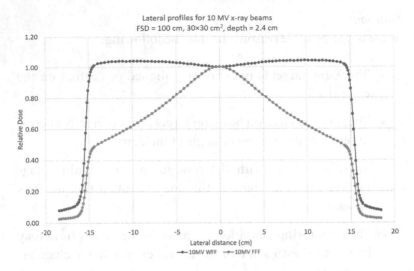

FIGURE 2.8   Depth dose curves and lateral profiles of 10 MV photon beams with (WFF) and without a (FFF) flattening filter.

- Higher dose rate for beam delivery, e.g., 2400 monitor units per minute for 10 FFF X-ray beams vs 600 monitor units per minute for 10 MV WFF X-ray beam on the Varian TrueBeam linear accelerators.

b. The flattening filter is made of a high atomic number metal, typically copper, tungsten or steel, which is cone-shaped. This design is used to attenuate the X-ray beam that exits the target. This beam is forward-peaked and the filter is used to create a flatter beam profile at depth in the patient or a water phantom. The linear accelerator manufacturers typically optimise the thickness and design of the filter so that the profile is flattest at a depth of 10 cm in the water phantom.

The flattening filter preferentially absorbs the lower energy photons more than the higher energy photons. This leads to X-ray beam hardening of the spectrum. However, because the flattening filter is thicker on the central axis, there is more beam hardening on the central axis and less as you move off axis. In addition, the photon interactions within flattening filter create secondary photon scatter which also has lower photon energy. This means that a FFF X-ray beam has a lower average photon beam energy compared to the WFF X-ray beam.

This means that the PDD for the FFF X-ray beam is slightly less penetrating as compared to the WFF X-ray beam. This also means that the surface dose is higher and the depth of maximum decreases a little for the FFF X-ray beam. For the beam profiles with the FFF X-ray beams, there is less variation of the profile shape with depth as there is less energy variations across the field.

c. The main application for FFF beams is with stereotactic ablative radiotherapy (SABR) and stereotactic radiosurgery (SRS) treatments. As these are typically high dose per treatment fraction and small field sizes, the use of FFF X-ray beams means that the treatment delivery is typically much quicker.

2.12 MAYNEORD FACTOR

Problem:
The PDD for a 15×15 cm field size, 10 cm depth and 80 cm source-to-surface distance (SSD) is 58.6 for a Co-60 beam. By using the Mayneord F factor approximation, find the PDD for the same field size and depth for a 100 cm SSD.

(Given $d_m = 0.5$ cm for ^{60}Co gamma-rays and

$$F = \left(\frac{f_2 + d_m}{f_1 + d_m} \right)^2 \times \left(\frac{f_1 + d}{f_2 + d} \right)^2)$$

Solution:

$$F = \left(\frac{f_2 + d_m}{f_1 + d_m} \right)^2 \times \left(\frac{f_1 + d}{f_2 + d} \right)^2$$

$$= \left(\frac{100 + 0.5}{80 + 0.5} \right)^2 \times \left(\frac{80 + 10}{100 + 10} \right)^2$$

$$= 1.5586 \times 0.6694$$

$$= 1.043$$

$$\therefore \frac{P(10,15,100)}{P(10,15,80)} = 1.043$$

Therefore, the desired percent depth dose is:

$$P(10,15,100) = P(10,15,80) \times 1.043$$

$$= 58.6 \times 1.043$$

$$= 61.1$$

2.13 OUTPUT FACTOR

Problem:

a. Explain the term 'output factor'.

b. Output factors are usually tabulated against equivalent square field size.

    i. What is equivalent field size for a rectangular field? How can it be determined?

    ii. Give an example by using a $12 \times 7 \, \text{cm}^2$ rectangular field.

c. Sketch a plot showing the relationship between output factor and equivalent square field size.

d. Explain the 'collimator exchange effect'.

Solution:

a. A linear accelerator is typically calibrated by measuring the ionisation per monitor units at the reference depth and field size which is usually $10 \times 10 \, \text{cm}^2$. As the field size changes, the radiation output changes. The ratio of output for a particular field size with the output for $10 \times 10 \, \text{cm}^2$ field at a reference depth is known as output factor.

b.

   i. For a rectangular field, an equivalent square field size is the field that would have the same output and depth dose characteristics. It can be determined by using the relationship: $S = (ab)/(a+b)$, where S is the side of the equivalent square, and a and b are the sides of the rectangular field.

   ii. For $12 \times 7 \, \text{cm}^2$ rectangular field, the equivalent field size is:

$$S = \frac{2 \times 2 \times 7}{12 + 7} = 8.8 \text{ cm}$$

c. See Figure 2.9.

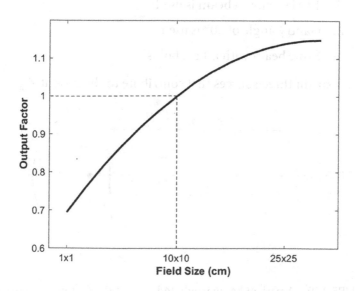

FIGURE 2.9 Relationship between output factor and equivalent field size.

d. The collimator exchange effect refers to the fact that the output factor can differ in rectangular fields dependent on which set of jaws is used to define short and long side of the field. Typically, the upper jaws, which are sitting closer to the target and the monitor chambers, affect the dose output more.

2.14 PDD AND SURFACE DOSE

Problem:

a. Figure 2.10 illustrates a patient being irradiated by a 6 MV photon beam from a distance, x cm source to SSD and zero gantry angle. Sketch a depth dose curve for photon beam as illustrated in Figure 2.10. Label the positions of Z_{max} on the curve.

b. On the same graph, sketch depth dose curves for the following conditions on your answer in (a):

   i. 10 MV photon beam is used

   ii. Gantry angle of 30° is used

   iii. Same beam with a 1 cm bolus

c. Explain three sources that contribute to the dose at Z_{ent}.

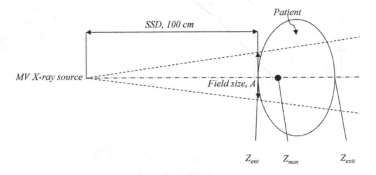

FIGURE 2.10 A patient being irradiated by a 6 MV photon beam from a distance, x cm source-to-surface distance (SSD) and zero gantry angle. (Abbreviations: Z_{en}: depth of entrance dose; Z_{max}: depth of maximum dose; Z_{exit}: depth of exit dose).

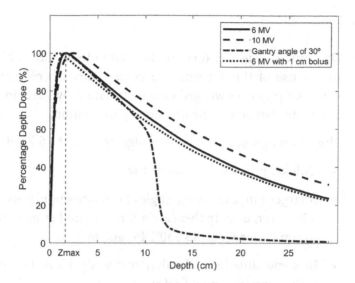

FIGURE 2.11 Depth dose curves for 6 MV photon, irradiation at 30° gantry angle and irradiation with a 1 cm bolus.

Solution:

Answers for (a) and (b) (Figure 2.11):

c. Photons scattered from the collimators, flattening filter and air, photons backscattered and secondary electron from the patient, high energy electrons produced by photons interactions in air and any shielding structures in the vicinity of the patient.

2.15 DYNAMIC WEDGE

Problem:

Dynamic wedges are widely available on linear accelerators and have largely replaced the older physical wedges which are typically made of high Z materials.

a. What is a dynamic wedge?

b. Explain the advantages of the dynamic wedge delivery to patients over the older fixed wedges.

Solution:

a. Dynamic wedge is a form of dose-rate modulation which makes use of the dynamic movement pairs of collimator jaws that produces wedged dose distribution by sweeping a collimator jaw across the field during irradiation.

b. The advantages of the dynamic wedges include the following:

- Quicker process for treatment setup.

- A larger range of wedge angles for treatment is possible – however, usually there are a limited number of angles commissioned, e.g., 15°, 30°, 45° and 60°.

- In some situations, the dynamic wedges can be used with larger treatment field sizes.

- Much easier to use for the treatment staff as there is less need to go in and out of the treatment room – this leads to faster treatment time for patients.

- A reduction in work, health and safety issues for both radiation workers and patients. Some of the large angle fixed wedges are very heavy which can be difficult to insert into the collimator.

- Less radiation safety issues with neutron activation in fixed wedges when used with X-ray beams >10 MV.

CHAPTER 3

Electron Beam Physics

3.1 R_P, R_{MAX} AND R_{50}

Problem:
Megavoltage electron beams represent an important treatment modality in modern radiotherapy, often providing a unique option in treatment of superficial tumours. Define the following for an electron beam:

a. Practical range (R_p)

b. Maximum range (R_{max})

c. R_{50}

Solution:

a. Practical range (R_p) – Depth at which the tangent plotted through the steepest section of the electron depth dose curve intersects with the extrapolation line of the background due to bremsstrahlung.

DOI: 10.1201/9780429159466-3

37

b. Maximum range (R_{max}) – Depth at which extrapolation of the tail of the central axis depth dose curve meets the bremsstrahlung background.

c. R_{50} – Depth of the electron depth dose curve at which percentage depth dose attains a value of 50%. This is also a very important depth as it defines the nominal electron energy as E (in MeV) = 2.33 × R_{50} (in cm).

3.2 STOPPING POWERS FOR ELECTRONS

Problem:
Explain briefly the two types of stopping powers for electrons.

Solution:
There are two types of stopping powers: the collision stopping power S_{col} and radiative stopping power S_{rad}.

The collision stopping power is due to Coulombic interactions between the incident electron and the orbital electrons in the atoms of the medium. As a result of the Coulombic interactions, the incident electrons transfer their kinetic energy by the processes of ionisation and excitation of the orbital electrons.

The radiative stopping power is due to the Coulombic interactions between the incident electron and the nucleus of the atoms in the medium. This interaction causes deceleration of the incident electron, and the energy is transferred to the atom which results in the generation of an emitted photon. This process is also known as bremsstrahlung.

3.3 ELECTRON BEAM ISODOSES

Problem:
Consider an electron beam of 16 MeV. Estimate the following:

a. Surface dose

b. Depth of 90% isodose

c. Depth of 80% isodose

d. Depth of 50% isodose

e. Practical range (R_p)

Solution:
A number of rules of thumb can be used to estimate parameters of electron beam isodose:

a. Surface dose = 70% + E = 86%

b. Depth of 90% isodose = $E/4$ = 4 cm

c. Depth of 80% isodose = $E/3$ = 5.3 cm

d. Depth of 50% isodose = $E/2.33$ = 6.9 cm

e. Practical range (R_p) = $E/2$ = 8 cm

Note: Rules of thumb for depth dose prediction are not perfect, but they do provide a valuable tool for predicting appropriate energies to avoid critical structures and provide adequate coverage to your treatment volume. For quantitative assessment, true electron depth dose curves should be referenced.

3.4 ELECTRON BEAM DEPTH DOSE CURVE

Problem:
Figure 3.1 shows a typical electron beam percentage depth dose curve. Indicate the parameters R_q, R_p, R_{50} and R_{max} on the curve.

Solution:
See Figure 3.2.

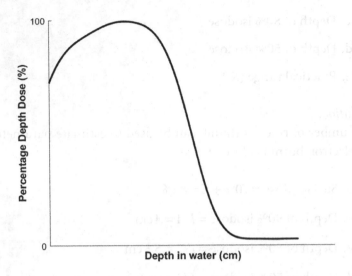

FIGURE 3.1   A typical electron beam percentage depth dose curve.

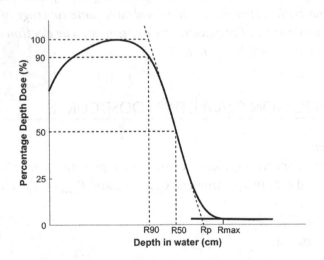

FIGURE 3.2   A typical electron beam percentage depth dose curve with indications of parameters R_{90}, R_{50}, R_p and R_{max}.

3.5 ELECTRON RADIOTHERAPY WITH BOLUS

Problem:

A lesion of 3 cm depth is treated with the 12 MeV electron beam. PDD for this beam is given in the table below. One centimetre of bolus is placed on the skin. The prescribed dose is 200 cGy to the distal edge of the lesion.

Depth (cm)	0	1	2	3	4	5	6
PDD (%)	90	95	98	100	80	40	5

a. Explain the advantage of electron beams compared with photon beams, from the physics perspective.

b. List three functions of the bolus in radiotherapy using electron beam.

c. What is the maximum skin dose?

d. What is the skin dose (in cGy) with the bolus in place?

Solution:

a. Electron deposits high dose at shallow depth and produces fast dose fall-off beyond the maximum dose, resulting in almost no dose beyond certain distance. This is advantageous to limit dose to organ at risk located beyond the treated lesion. This is in contrast with photon beam that contributes relatively low dose at shallow depth, with significant amount of dose deposited at region beyond the depth of maximum dose.

b. In radiotherapy using electron beam, a bolus can be used to (i) increase the surface dose, (ii) flatten out the uneven patient surface and (iii) reduce the dose penetration of the electrons in parts of the field.

c. The prescribed dose is 200 cGy to the distal edge of the lesion, i.e. 200 cGy to the depth of 4 cm (1 cm bolus + 3 cm

lesion depth). This depth is covered by 80% isodose. Therefore, maximum skin dose is $\dfrac{200 \text{ cGy}}{80\%} = 250 \text{ cGy}$.

d. Skin dose is $0.95 \times 250 \text{ cGy} = 238 \text{ cGy}$.

3.6 ELECTRON BEAM ENERGY SPECIFICATION

Problem:
The energy distribution of an electron beam varies with depth in a
phantom. There is no single energy parameter that can fully characterise the electron beam. State four parameters that can be used
to describe an electron beam.

Solution:
The most probable energy $E_{p,0}$ on the phantom surface is empiri-
cally related to the practical range R_p in water as follows:

$$E_{p,0} = 0.22 + 1.98\, R_p + 0.0025\, R_p^2$$

where $E_{p,0}$ is in MeV and R_p in cm.

The mean electron energy $\bar{E}_0$ at the phantom surface is related
to the half-value depth R_{50} as follows:

$$\bar{E}_0 = CR_{50}$$

where $C = 2.33$ MeV/cm for water.

The depth R_{50} is calculated from the measured value of I_{50}, the
depth at which the ionisation curve falls to 50% its maximum, by

$$R_{50} = 1.029\, I_{50} - 0.06 \ (\text{cm}) \ \ (\text{for } 2 \le I_{50} \le 10 \text{ cm})$$

$$R_{50} = 1.059\, I_{50} - 0.37 \ (\text{cm}) \ \ (\text{for } I_{50} > 10 \text{ cm})$$

$\bar{E}_z$, the mean energy at a depth z in a water phantom is related to
the practical range R_p by the Harder equation as follows:

$$\bar{E}_z = \bar{E}_0 \left(1 - \frac{z}{R_p} \right)$$

3.7 BREMSSTRAHLUNG CONTAMINATION

Problem:
In a linear accelerator operating in electron mode, the X-ray target is removed from the electron beam path. Nevertheless, bremsstrahlung contamination still exists in electron beam irradiation.

    a. What causes bremsstrahlung contamination in electron beam irradiation?

    b. With the aid of a diagram, described how bremsstrahlung contamination is featured on an electron depth dose curve.

    c. What factors affect the fraction of bremsstrahlung X-ray generated in electron beam irradiation?

Solution:

    a. Bremsstrahlung contamination is produced in the head of the accelerator through interaction with materials acting as 'targets' such as the waveguide window, scattering foils, ionisation chamber, and collimators, present in the path of the electron beam. Using low Z materials in the scattering foils can reduce this unwanted contamination. Bremsstrahlung is also generated in the air between the accelerator window and the patient, and in the irradiated medium.

    b. Bremsstrahlung contamination is featured as 'a tail' extending beyond the range of the most energetic electron on an electron depth dose curve (Figure 3.3).

    c. Factors affecting the fraction of bremsstrahlung radiation generated are (i) electron energy – fraction of bremsstrahlung radiation generated increases with increasing electron energy; (ii) materials of collimation system – rate of bremsstrahlung production is proportional to the atomic number of the collimation system; and (iii) thickness of the scattering foils – fraction of bremsstrahlung radiation generated increases with increasing thickness of scattering foils.

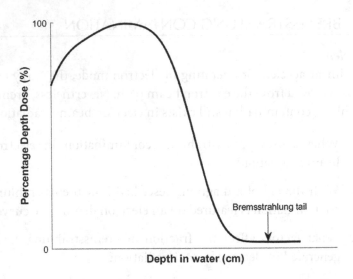

FIGURE 3.3   Bremsstrahlung contamination on an electron depth dose curve.

3.8 ELECTRON BEAM SHAPING

Problem:
Explain the different methods for collimating and beam shaping electron beams.

Solution:
The electron beams when emitted from the collimator of the linear accelerator have to travel a reasonable distance until they are incident on the patient. As the electrons readily scatter as they are travelling through air, it is important that they are collimated to ensure the beam is directed to the area of interest on the patient. Otherwise, radiation doses will be scattered over a large area of the patient.

The primary collimation of the electron beams is via the 'electron applicators' which are attached to the end of the collimator. These are large metal devices and extend to a distance typically of SSD of 95 cm. This means that there is a small air gap between the end of the applicator and the patient surface.

The patient-specific collimation is achieved by using a metal electron cut-out which is fixed into the end of the applicator. This is made of a low-melting alloy (LMA) with an aperture that has a shape that is required for the patient. This means that the field edge will be well defined for the treatment area.

However, the air gap of 5 cm still does mean that there is some penumbral broadening. A sharper beam edge can be achieved by placing a high Z material cut-out, typically made from lead, on the patient surface. One scenario for this is the use of electron treatments near the eye where we wish to reduce the dose to the lens as low as possibly achievable.

3.9 ELECTRON CUT-OUT

Problem:
A skin lesion on the chest wall is to be treated to a depth of 2 cm with electrons. The end of the applicator is 5 cm from the patient's skin. Figure 3.4 shows the electron cut-out used in this treatment.

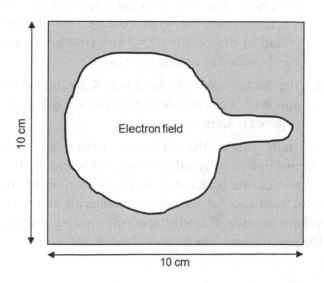

FIGURE 3.4   A cut-out used for radiotherapy using electron beam.

a. Estimate the electron beam energy to treat to 2 cm with 90% isodose level.

b. From your answer in (a), what is the minimum thickness required of the cut-out made from:

   i. Lead

   ii. Cerrobend or LMA

c. What is the problem of using a cut-out as shown in the figure above? Explain your answer.

Solution:

a. To estimate the 90% isodose level, use the rule of thumb of $E/4$, where E is the electron energy. For a lesion of 2 cm depth, the energy required would be 8 MeV to obtain 90% isodose level (or 90% at 1.5 cm).

b.

   i. The rule of thumb for calculating lead thickness in mm is $E/2$, where E is the electron energy. Thus, the thickness of lead in this case is $8/2 = 4$ mm (another mm of lead may be added as a safety margin).

   ii. The thickness of Cerrobend is 20% greater than that of pure lead. Thus, in this case the thickness of Cerrobend required is 4.8 mm.

c. The right part of the field is too narrow and would be affected by lack of lateral electron equilibrium. Such narrow field will cause lateral electron disequilibrium where there is an imbalance between electrons entering and exiting the volume laterally. This will impact the absolute dose as well as the penetration of the beam at this volume.

Note: The rule of thumb for question (a) can only be used as an estimation based on the lesion depth. It does not consider the treated area up to the skin, which may require bolus, and thus different electron energy to provide adequate coverage. For clinical quantitative assessment, actual electron depth dose curves should be referenced.

3.10 ISODOSE CURVES

Problem:
An isodose curve is a line passing through points of equal dose.

a. Explain the characteristics of typical electron beam isodose curves.

b. How is the physical penumbra of an electron beam defined?

c. How does the low- and high-value isodose lines change with increasing air gap between the patient and the end of the applicator?

Solution:

a. For low electron energies, the low-value isodose curves (<50%) bulge out. On the high-dose side electron beams exhibit a lateral constriction of the higher value isodose curves (80%). As this 'bulging' in both directions increases with depth as the electrons increase the lateral component of the travel, the effective penumbra between high- and low-isodose lines increases with depth.

b. The physical penumbra of an electron beam may be defined as the distance between two specified isodose curves at a specified depth. The ICRU has recommended that the 80% and 20% isodose lines be used in the determination of the physical penumbra, and that the specified depth of measurement be $R_{85/2}$, where R_{85} is the depth of the 85% dose level beyond the depth of maximum dose on the electron beam central axis.

c. The low-value isodose lines diverge with increasing air gap between the patient and the end of the applicator, while the high-value isodose lines converge towards the central axis. This implies that the penumbra will increase if the distance from the applicator increases.

CHAPTER 4

Treatment Planning for External Beam Radiotherapy

4.1 TARGET VOLUMES AND MARGINS (1)

Problem:

a. Label the planning volumes (i) to (v) in Figure 4.1 based on the International Commission on Radiation Units and Measurement (ICRU) Report No. 50.

b. Define the volumes (iii) and (iv).

c. An additional volume is added between volumes (iv) and (v) in ICRU Report 62. Name this volume.

d. What is meant by the term the 'conformity index' (CI)?

DOI: 10.1201/9780429159466-4

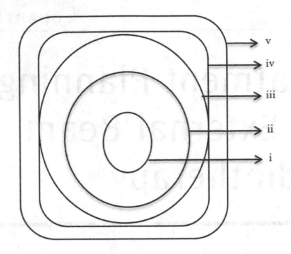

FIGURE 4.1   Definitions of target volumes as recommended by ICRU Report No. 50.

Solution:

a.

   i. Gross tumour volume (GTV)

   ii. Clinical target volume (CTV)

   iii. Planning target volume (PTV)

   iv. Treated volume (TV)

   v. Irradiated volume (IV)

b. Volume (iv) is the treated volume (TV). It is a volume enclosed by an isodose surface, selected and specified by the treating oncologist as being appropriate to achieve the purpose of treatment (e.g., cure or palliation).

   Volume (v) is the irradiated volume (IV). It is a tissue volume that receives a dose that is considered significant in relation to normal tissue tolerance.

c. Internal target volume (ITV). The ICRU Report 62 distinguishes between two concepts that contribute to the margin required to design a PTV to ensure that the CTV is covered adequately under all circumstances. The set-up margin takes into account all uncertainties that occur when trying to bring the target (typically represented by a reference point) into the correct location. Examples are lasers, field size, couch sag and of course patient positioning. The internal margin refers to all changes that can affect a target volume as it relates to this set-up. Organ motion due to breathing is the most commonly used example but also organ filling (e.g., bladder) or deformation is covered in the ITV concept.

d. Conformity Index is an important metric for determining how tightly the prescription dose is conforming to the target. The ICRU Report 62 defines conformity index as $CI = \dfrac{TV}{PTV}$, where TV is the treated volume enclosed by a given isodose surface (e.g., 95%) and PTV is the planning target volume.

4.2 TARGET VOLUMES AND MARGINS (2)

Problem:
A treatment plan for prostate cancer is produced using computed tomography (CT) images. An oncologist produces five volumes, as shown in Figure 4.2.

a. Which of the above volume(s) account for:

  i. Day-to-day variations in the location of prostate.

  ii. Gross prostate gland as identifiable from the CT scan.

  iii. Subclinical microscopic cancerous cells of the prostate.

  iv. Organs at risk during prostate radiotherapy.

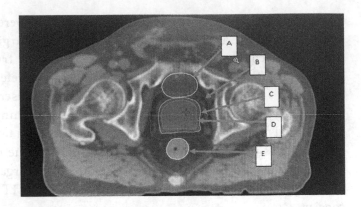

FIGURE 4.2 A treatment plan for prostate cancer is produced using computed tomography (CT) images.

b. The data presented to an oncologist during evaluation of treatment plan consist of (i) an isodose distribution on the CT slices and (ii) dose volume histograms (DVH) for the target volume and organs at risk. Describe specific information that can be obtained from:

i. Isodose distribution

ii. DVH

Solution:

a. (i) B, (ii) D, (iii) C and (iv) A and E

b. Specific information that can be obtained from:

i. Isodose distribution – Dose coverage or distributions on the target volume(s) and organs at risk (OAR), homogeneity of dose distribution, spatial distribution/location of hot or cold spots.

ii. DVH – Dose summary or statistics of the target volume(s) and OAR, e.g., minimum, maximum, mean and modal doses as well as some dose parameters such as D_{95} (dose received by at least 95% of the volume) and V_{20} (volume irradiated to at least 20% of the prescribed dose).

4.3 CT NUMBER CALIBRATION (1)

Problem:

Sketch a graph of CT number versus relative electron density for a kVCT simulator with range from air to hard bone. Place some appropriate data points on the sketched graph for the lung, muscle and hard bone and soft bone. Describe how much lung's relative electron density can change with inspiration and expiration and how we can minimise this problem when CT scanning a patient.

Solution:

A graph of CT number versus relative electron density for a CT simulator is shown in Figure 4.3. The relative electron density of the lung can change by 0.2 with inspiration and expiration. Children also typically have higher lung densities and there are often differences between anterior and posterior aspects of the lung. Breath-hold technique and the use of 4DCT can help to control this problem.

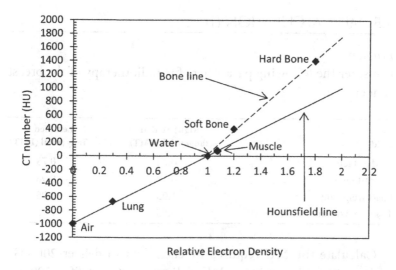

FIGURE 4.3   Graph of CT number versus relative electron density for a CT simulator.

4.4 CT NUMBER CALIBRATION (2)

Problem:
Compare the graph in Problem 4.3 to the CT number versus relative electron density curve obtained from a megavoltage computed tomography (MVCT) of a tomotherapy unit. Explain the reason for the difference.

Solution:
As megavoltage photons mostly interact through Compton interaction which is independent of the atomic number of the material, the CT number versus relative electron density relationship for MVCT is linear, without the kink in the bone line. The kink in the bone line in the CT calibration curve obtained using a kV CT is due to the predominant photoelectric absorption of high-Z material such as the bone, and due to the large difference in the mass absorption coefficients of the materials with different effective atomic number, Z. In the megavoltage energy range, the Compton effect is more dominant, and the mass absorption coefficients of the different Z materials are very similar.

4.5 MU CALCULATION (1)

Problem:
Consider the following parameters for radiotherapy of the breast cancer:

Field	Right Medial Tangential (RMT)	Left Medial Tangential (LMT)
Wedge factor	0.74	0.75
Output factor	0.98	0.98
Field weighting	1.05	0.95
Tissue maximum ratio (TMR)	0.85	0.85

Calculate the MU required by each field to deliver 200 cGy to isocentre. (Assuming 1 MU delivers 1 cGy at SSD 100 cm, 10 cm × 10 cm field size and machine is calibrated isocentrically.)

Solution:
Dose delivered by right medial tangential,

$$D_{RMT} = D_{iso} \times \frac{Weighting_{RMT}}{Weighting_{total}}$$

$$= 105 \text{ cGy}$$

Dose delivered by right lateral tangential,

$$D_{RLT} = D_{iso} \times \frac{Weighting_{RLT}}{Weighting_{total}}$$

$$= 95 \text{ cGy}$$

Thus, monitor units,

$$MU_{RMT} = \frac{D_{RMT}}{\dfrac{1 \text{ cGy}}{MU} \times Output\ factor \times Wedge\ factor \times TMR}$$

$$= \frac{105}{\dfrac{1 \text{ cGy}}{MU} \times 0.98 \times 0.74 \times 0.85}$$

$$\approx 170$$

$$MU_{LMT} = \frac{D_{LMT}}{\dfrac{1 \text{ cGy}}{MU} \times Output\ factor \times Wedge\ factor \times TMR}$$

$$= \frac{95}{\dfrac{1 \text{ cGy}}{MU} \times 0.98 \times 0.75 \times 0.85}$$

$$\approx 152$$

4.6 MU CALCULATION (2)

Problem:
Calculate the MU per fraction which needs to be set to deliver a dose of 60 Gy in 30 fractions to a position in the tumour. The tumour is at the isocentre and hence this is a SAD treatment with the tumour at 10 cm depth. A single 18 MV beam of field size (*S*) 20 cm × 20 cm is used. The TMR for this field at 10 cm depth is 0.898, and the output factor (OF) for this field size at 100 cm SSD is 1.07. The d_{max} is 3.3 cm.

Solution:
Summary of information:
   Prescribed dose, $D = 60$ Gy in 30 fractions → 200 cGy per fraction
   SAD treatment is employed.
   $d = 10$ cm
   $E = 18$ MV
   $S = 20$ cm × 20 cm
   TMR (10,20) = 0.898
   OF (20 cm) = 1.07
   $d_{max} = 3.3$ cm

F Calibration: $\left(\dfrac{D}{MU} \right)_{std.SSD,10\times10,water} = 1cGy/MU$

Calculation:

Inverse square factor, $INVSQ = \left(\dfrac{f + d_{max}}{f} \right)^2 = \left(\dfrac{100 + 3.3}{100} \right)^2$

$$= 1.067$$

$$MU\ required = \frac{D}{\left(\dfrac{D}{MU} \right)_{SSD} \times INVSQ \times OF \times TMR}$$

$$= \frac{200}{(1)_{SSD} \times 1.067 \times 1.07 \times 0.898} = 195$$

4.7 ISODOSE CURVE

Problem:

a. What are 'isodose curves' and how are they produced?

b. Contrast the advantages and disadvantages of isodose curve display and colour wash display.

Solution:

a. Isodose curves are defined as lines that join points of equal dose. They offer a planar representation of the dose distribution. They can be measured directly using a beam scanning device in a water phantom. They can be calculated from percentage depth dose and beam profile data. They can be adopted from an atlas or a catalogue for isodose curves.

b. Isodose curves can be shown either as absolute dose (in Gy) or relative dose (as % of the prescription dose). It is important to check what is used. Isodose curves provide a numerical dose value for selected points while a colour wash display gives a visual representation of dose that is familiar to many clinicians, e.g., from nuclear medicine image displays. It can be made semi-quantitative by specifying the minimum and maximum dose on display.

4.8 WEDGE (1)

Problem:
Explain briefly three types of wedge filter.

Solution:
Three types of wedge filters are currently in use: manual, motorised and dynamic. Physical wedges are angled pieces of lead or steel that are placed in the beam to produce a gradient in radiation intensity. Manual intervention is required to place the physical wedges on the treatment unit's collimator assembly. A motorised

wedge is a similar device, a physical wedge integrated into the head of the unit and controlled remotely. A dynamic wedge produces the same wedged intensity gradient by having one jaw close gradually while the beam is on.

4.9 WEDGE (2)

Problem:
With the aid of sketches showing the shape of a wedge, dose profile and isodose level shape, describe what a 30° wedge does.

Solution:
Wedge angle is defined as the angle between the 50% isodose line and the perpendicular to the beam central axis. Thus, a 30° wedge will tilt the 50% isodose line by 30° as illustrated in Figure 4.4.

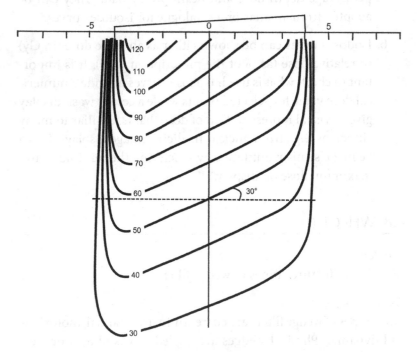

FIGURE 4.4 Figure showing the tilted isodose profiles with the presence of a 30° wedge.

4.10 EFFECTIVE DEPTH CALCULATION

Problem:

Despite the wide availability of computerised treatment planning systems and monitor units check programmes, it is still useful to be able to get a 'feel' for dose distributions by performing hand calculations. A 6 MV X-ray beam (field size 10 cm × 10 cm at 100 cm SSD) penetrates through the heads of femur bone of a prostate patient. The tissue prior to the head of femur is 5 cm of fat, then the head of femur thickness is 5 cm, and then the beam traverses 6 cm of muscle and 3 cm of the prostate tissue before reaching the dose point of interest. The electron densities of various tissues are as follows:

Tissue	Electron Density, ρ_e^w (g/cm³)
Fat	0.95
Head of femur	1.2
Muscle	1.0

Given that the percentage depth doses at 19 and 20 cm are 41% and 39%, respectively,

a. Calculate the effective depth to the point of interest (ignore surface curvature).

b. Calculate the correction factor which would be applied to the percentage depth dose value using effective depth correction.

c. Calculate the new effective percentage depth doses.

Solution:

a. Effective depth, $d' = \sum_{i=1}^{4} d_i \rho_i = (5 \times 0.95) + (5 \times 1.2) + (9 \times 1.0)$

$= 19.75$ cm

b. Geometric depth, $d = 5 + 5 + 6 + 3 = 19\,\text{cm}$
%D (19 cm) = 41%, %D (20 cm) = 39%
Through mathematical interpolation,
%D (19.75 cm) = 39.5%

c. Correction factor, $CF = \dfrac{D(d')}{D(d)}\left(\dfrac{f+d'}{f+d}\right)^2 = \dfrac{39.5}{41}\left(\dfrac{100+19.75}{100+19}\right)^2$

$= 0.9756$

Effective percentage depth dose, $\%D(d')_{\text{inhom}} = \%D(d)_{\text{water}} \times CF$

$$\%D(d')_{\text{inhom}} = \%D(d)_{\text{water}} \times CF = 41\% \times 0.9756 = 40\%$$

4.11 ICRU CRITERIA (1)

Problem:
There are several ways how a prescription to the target can be specified based on the ICRU criteria. A lot of clinical experience is based on the target dose specified and recorded at what is called the ICRU reference point.

a. List four general criteria for this point.

b. Many modern radiotherapy delivery techniques are difficult to prescribe using this concept and ICRU Report 83 which is on IMRT describes a system that is based on prescription to isodose coverage. If dose is prescribed to a covering isodose whose dose level is closely related to the traditional prescription to a reference point?

c. Why is the minimum dose not recommended for IMRT dose specification to the PTV?

d. Define 'near minimum' and 'near maximum' dose according to ICRU Report 83.

Solution:

a. (i) The dose to the point should be clinically relevant; (ii) the point should be easy to define in a clear and unambiguous way; (iii) the point should be selected so that the dose can be accurately determined; and (iv) the point should be in a region where there is no steep dose gradient.

b. The median dose as it is usually nearest to the dose value of the more traditional ICRU reference point.

c. This is because the minimum dose represents the dose to a single point likely at the boundary of the PTV and is far too susceptible to errors in delineation.

d. 'Near minimum' dose is dose to 98% of the volume (D_{98}) and 'near maximum' dose is dose to 2% of the volume (D_2).

Notes: The answers in (c) are based on the concepts that take into account typical target volumes and associated contouring and dose calculation accuracy. They may need to be modified for small and large targets.

4.12 ICRU CRITERIA (2)

Problem:
Figure 4.5 shows a treatment plan for a breast cancer patient to be treated with 6 MV photon beam using half-beam block technique.

a. Which of the points (Points 1 to 3) is most appropriate as the ICRU reference point? Give reasons for your choice of reference point.

b. State why the other two points should not be chosen as the ICRU prescribing point.

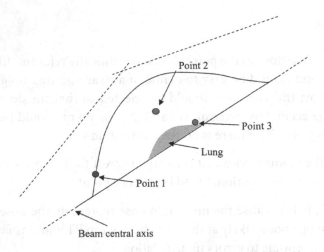

FIGURE 4.5 A treatment plan with two half-beam block tangential beams for radiotherapy of breast cancer.

Solution:

a. Point 2 should be the ICRU prescription point. It meets the criteria of an ICRU reference point:

- It is a point where the dose is clinically relevant which can be a good representative of the dose to the target volume.

- The dose at this point can be clearly defined and unambiguous.

- The point is located at the centre of the PTV, and highly likely at a region with no steep dose gradients.

b. Point 1 is located at the skin while Point 2 is located at the interface between the tissue and the lung. Both points do not meet a number of criteria as an ICRU reference point. Both points are not located at the centre of the target volume. They are located at interfaces with high inhomogeneity, where steep dose gradients are expected. Doses at these points may be unclear, have greater uncertainty and not reproducible.

4.13 BREAST TREATMENT PLAN

Problem:

Figure 4.6 shows a radiotherapy treatment plan with isodose values for breast cancer using parallel opposed tangential fields (6 MV) using an isocentric technique.

a. What is the advantage of isocentric techniques over fixed SSD technique for radiotherapy treatment?

b. After a review, the plan is found to be unsatisfactory. Suggest a few reasons.

c. Draw the orientations of wedges to improve the dose distribution of the plan.

Solution:

a. For isocentric technique, patient's position remains the same throughout treatment, whereas for fixed SSD, treatment position changes for every treatment field. Radiation therapists have to change the positions for every treatment field which is more susceptible to set up errors.

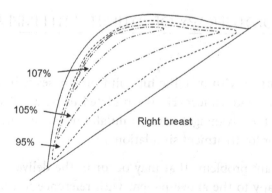

FIGURE 4.6   Isodose distribution of a breast radiotherapy plan.

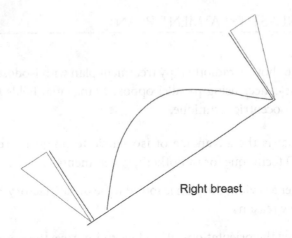

FIGURE 4.7 Correct orientation of wedges for improving dose distribution in breast radiotherapy.

b. Dose not homogenously distributed throughout the breast (target volume); being relatively hot at the top area (surrounding nipple), and relatively cold downwards. There is the hot area where isodose is more than 107% and not all target volumes are encompassed by 95% as recommended by the ICRU.

c. See Figure 4.7.

4.14 RADIOTHERAPY OF PATIENT WITH IMPLANT

Problem:

a. A patient with metallic hip joint prostheses will be treated for prostate cancer. He is simulated using a CT simulator. Give the advantages of CT simulator over conventional simulator for treatment simulations.

b. Explain problems that may occur in the delivery of radiotherapy to the above patient with reference to the metallic implant.

c. It has been decided to track the motion of the prostate gland real time during radiotherapy. Suggest two non-ionising IGRT modalities that can be used.

Solution:

a. Advantages of CT simulator over conventional simulator for treatment simulations:

- A CT simulator provides the ability to localise very precisely the target volumes and critical structures. It allows for accurate identification and delineation of these structures directly onto the CT data set.

- Patient is usually not required to stay after the scanning has taken place.

- It allows the user to generate digitally reconstructed radiographs and beam's eye views even for beam geometries that were previously impossible to simulate conventionally.

b. Problems that may occur in the delivery of radiotherapy to the above patient with reference to the metallic implant:

- Image artefacts caused by the metallic prostheses will cause difficulties in delineation of structures and can cause error in dose calculation.

- Shadowing effect (attenuation of X-ray) may occur on the tissues beyond the metallic implant because of the higher density and atomic number.

- The dose immediately adjacent to the metallic prostheses will be enhanced by the electron flux generated within the metal both penetrating the adhesive and the bone itself. Adhesives may lose their adhesion capability under radiation.

c. Any two of the answers: ultrasound, implanted transponders with electromagnetic array, online imaging using X-ray.

4.15 BOLUS VERSUS COMPENSATING FILTER

Problem:
Figure 4.8 shows two different treatment apparatuses used during radiotherapy of squamous cell carcinoma of the nasal septum using high-energy photon beam.

a. Name the apparatus (i) and (ii).

b. What are the common materials used to construct (i) and (ii)?

c. Describe two differences on the use of (i) and (ii) for radiotherapy.

d. List two disadvantages of using (ii) compared to (i).

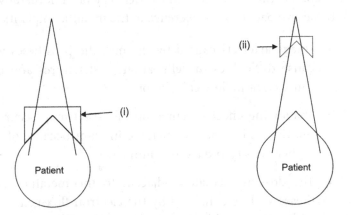

FIGURE 4.8 Two different types of beam treatment apparatus used during radiotherapy.

Solution:

a. (i) Bolus or wax bolus; (ii) compensator or compensating filter.

b. (i) Wax; (ii) metal or LMA, e.g., Cerrobend.

c. A bolus is placed directly on the patient while a compensating filter is attached onto the head of the linear accelerator using a tray holder during radiotherapy. The use of a bolus does not provide skin sparing effect, while the use of a compensating filter provides skin sparing effect.

d. Two disadvantages of using a compensating filter in comparison to a bolus:

- More laborious and time-consuming.

- The resulting dose distribution cannot be readily calculated on most TPS without measurement under the compensator and additional beam data entry into the TPS.

4.16 SKIN CANCER TREATMENT ON THE NOSE

Problem:
Discuss the clinical and dosimetric issues that would be needed to be considered in treating a skin cancer on the side of the nose with kilovoltage X-ray beams. The PTV has been defined by the radiation oncologist to be approximately 2 cm in diameter. You should include any comments on verification of the dose delivered to the patient.

Solution:

- X-ray beam energy – as this is a skin cancer, it would be expected that the radiation oncologist would likely want to treat to several mm depth below the skin. Depending on the penetration required, the X-ray beam energy would typically range between 75 and 125 kVp (with corresponding HVLs of several mm Al).

- Beam collimation – there are two options: (i) the use of a 2 cm diameter applicator or (ii) a larger lead cut-out with 2 cm diameter hole to be used with a larger size applicator.

- Patient curvature – due to the curvature of the nose, the applicator should be directed straight towards the tumour volume and avoid angulation wherever possible.

- Bolus – tissue equivalent bolus material should not be placed on top of the PTV as the maximum dose for the kilovoltage X-ray beam occurs at the skin surface. However, bolus such as a wax mould can be placed around the nose region in order to provide optimal scatter conditions.

- Internal shielding – the use of internal metal shielding inside the nostril could be considered to reduce the dose to healthy tissues. However, it should be noted that there can be significant backscatter dose from the shielding which could be reduced by coating the internal shield with wax. Alternatively, the nostril could be packed with some form of bolus-like material, e.g., wet gauze or cotton to reduce dose perturbations.

- Patient dose calculations – the monitor units required for the patient need to be carefully determined using the appropriate applicator or cut-out factor. The presence of the bone/cartilage in the nose will cause some form of dose perturbation.

- *In-vivo* dose verification – one can make use of a suitable dosimeter to verify the dose to the patient. For kilovoltage X-ray beams, radiochromic film and TLDs are likely the best options due to the size, spatial resolution and energy response. Other dosimeters like diodes have a large energy response variation due to the higher Z of silicon and OSLDs have a thicker casing which can cause dose averaging.

CHAPTER 5

Radiation Dosimetry

5.1 BRAGG-GRAY CAVITY THEORY

Problem:
List the three conditions for which a detector can be considered to be a Bragg-Gray cavity.

Solution:
The three conditions for which a detector can be considered to be a Bragg-Gray cavity:

- The cavity of the detector must not disturb the fluence of charged particles that exists in the medium in the absence of the cavity. This means that the size of the cavity must be smaller than the range of the charged particles in the particular medium of interest. For most radiotherapy measurements, the suitable detectors are ionisation chambers which have a small cavity which contains air.

- The absorbed dose within the cavity is deposited entirely by the charged particles that are traversing across the cavity. There should not be any dose contribution due to photon interactions that occur within the cavity itself. If photon

DOI: 10.1201/9780429159466-5

interactions were to occur in the cavity, then the fluence of charged particles will be different as compared to the fluence of charged particles that occur just within the medium without the cavity present.

- Charged particle equilibrium must exist in the medium at the position of the detector but without the presence of the cavity in the medium.

5.2 CHARACTERISTICS OF IDEAL DOSIMETERS

Problem:
Discuss briefly the characteristics of an ideal dosimeter.

Solution:

- Accuracy – This is the ability of the detector to correctly indicate dose. Statements of accuracy must take into considerations the incidence of errors in detector performance and measurement technique. Such errors are split into two categories: random and systematic errors. Random errors are unpredicted events whose magnitude is reduced by taking the average of a number of measurements. Systematic errors are predictable, have consistent magnitude and are the result of subjective displays of detector output and subsequent observer interpretations, as well as from variations in detector and machine performance. This is the capability of the detector to produce usable signal for the given type and energy of radiation accuracy with respect to the incidence of systematic errors.

- Specificity – This is the capability of the detector to produce usable signal for the given type and energy of radiation.

- Sensitivity – This is the capability of the detector to produce a signal of sufficient magnitude for very small amounts of radiation intensity as well as for the larger intensity values.

- Energy resolution – This is the ability to distinguish between two or more closely lying energies.

- Response time – The duration of time the detector takes to form a signal after the arrival of the radiation.

- Signal duration – The capability of the detector to register a second radiation event while in the process of producing a signal as a result of the first radiation event.

- Precision – This can be defined as the reproducibility of the detector response under equivalent irradiation conditions. It is important the same reading is produced following the receipt of the same amount of radiation.

- Dose response – Over a defined energy range, it is helpful if the detector produces a signal that is proportional to the amount of incident radiation.

- Objectivity of the method of data display – The ability of the data to be displayed in a method that produces minimal subjective interpretation by the observer.

- Detector efficiency – This is split into two categories: absolute and intrinsic. Absolute detector efficiency is a measure of its ability to measure the total events emitted by the source of ionising radiation. Intrinsic efficiency reflects the ability of the detector to produce a signal for each quantum of radiation incident upon it.

5.3 PROPERTIES OF CYLINDRICAL IONISATION CHAMBERS

Problem:
List the recommended properties of a cylindrical ionisation chamber for the purpose of reference dose calibrations based on the recommendations of the IAEA TRS398 Code of Practice (CoP).

Solution:

- The ionisation chamber cavity volume should be between about 0.1 and 1 cm³.

- The air cavity should have an internal diameter not greater than around 7 mm and an internal length not greater than 25 mm.

- The air cavity should not be sealed; it should be designed so that it will equilibrate rapidly with the ambient temperature and air pressure.

- The chamber construction should be as homogeneous as possible.

- The walls of the ionisation chambers are usually made of graphite which means they have a good long-term stability and a uniform response.

- That walls of the ionisation chamber should not be made of particular plastic materials such as A-150 or nylon as their response varies according to the humidity of the air.

- The central electrode of the ionisation chamber is usually a different material such as aluminium.

The energy response of the chamber should not vary significantly particularly for low-energy X-ray beams. Note that this is evident from the range of air-kerma calibration factors for the chamber.

5.4 IONISATION CHAMBERS FOR ELECTRON BEAM DOSIMETRY

Problem:
According to the IAEA TRS398 CoP, why are plane-parallel ionisation chambers recommended for use in electron beam dosimetry? List the required design features for such ionisation chambers.

Solution:

Plane-parallel ionisation chambers are designed so that they min-imise scattering effects by the chambers. The thin front window samples the incident electron beam while the design of the side walls is such that there are only minimal electrons that enter the cavity from the side.

The required design features for plane-parallel chambers are as follows:

- The thickness of the front window should be no greater than $0.1 \, \text{g} \, \text{cm}^{-2}$ or 1 mm PMMA.

- Have a disc-shaped air cavity.

- The air cavity must be vented to the atmospheric conditions so that it can equilibrate quickly to the ambient temperature and air pressure.

- The ratio of the cavity diameter to the cavity height should be large typically equal to or greater than 5.

- The diameter of the collecting electrode should not exceed 20 mm – this is done to minimise any variations in the radial uniformity of the electron beam profiles.

- The air cavity height should not be more than 2 mm.

- The collecting electrode should be surrounded by a guard electrode with a width that is not smaller than 1.5 times the cavity height.

5.5 BASIC PRINCIPLE OF AN IONISATION CHAMBER

Problem:

With the aid of a diagram, describe briefly the basic principle of the operation of an ionisation chamber.

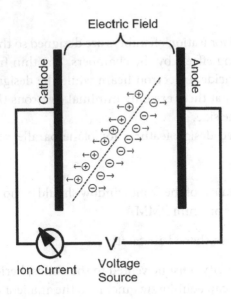

FIGURE 5.1 Schematic diagram on operational principle of an ionisation chamber.

Solution:

A schematic diagram on the operational principle of an ionisation chamber is shown in Figure 5.1.

An ionisation chamber operates based on the principle of measuring the number of ion pairs produced in a volume of air due to radiation. In the simplest arrangement, the ionisation chamber exists as two-electrode plates spaced apart in air. A large potential (100–400 V) is applied to the plates. Radiation dose is delivered by charged particles in excitation and ionisation events. Charged particles are either the radiation of interest themselves (e.g., electron and proton radiotherapy) or indirectly produced by non-charged radiation (e.g., photons and neutrons). When charged particles traverse between the plates, they ionise the air producing free negative electrons and positive ions. The positive- and negative-charged particles are then swept by the electric field between the plates towards the appropriate electrodes, producing a steady current flow in the external circuit, which can be measured by an electrometer. The important premise of an ionisation chamber is that each interaction of the charged particle produces exactly one ion pair and therefore allows for accurate quantification of dose.

5.6 GEIGER-MULLER (GM) COUNTER

Problem:
Geiger-Muller (GM) counter features a similar design as an ionisation chamber. Can a GM counter be used as a dosimeter? Explain your answer. List examples for GM use.

Solution:
A GM counter cannot be used as a dosimeter. This is because GM counter operates at a high-voltage region (800–1000 V). In this region, avalanche of charges occurs. The ion pairs generated by radiation gain such high energies in the accelerating voltage that they can produce new charges in secondary collisions. Amplification occurs. The signal generated is not proportional to the initial ionisation event as one or multiple ion pairs generated by the traversing charged particle produce the same signal: one 'count'. GM counters are suitable to be used in area monitoring, room monitoring and personnel monitoring for detecting the presence of ionising radiation.

5.7 DIODES

Problem:
Diodes are commonly used for relative dosimetry and *in vivo* dosimetry. List some advantages and disadvantages of diodes as dosimeters.

Solution:

Advantages	Disadvantages
• Direct reading/ fast readout/ real-time readout • Sensitive • Small size/volume • Waterproofing possible • Robust	• Angular-dependent • Temperature-sensitive • Sensitivity may change, periodic re-calibration necessary • Regular QA procedures need to be followed • Energy-dependent • Dose rate-dependent • Cables required

5.8 RADIOCHROMIC FILM DOSIMETRY

Problem:

a. Provide an overview of radiochromic film dosimetry.

b. Draw a typical dose-response curve for Gafchromic EBT3 film for the three colour channels red, green and blue? Comment on the use of the different colour channels.

c. What are the key aspects for a good QA programme for radiochromic film?

Solution:

a. Radiochromic films are sensitive to ionising radiation. These films have a monomer within a polymer matrix that is then coated onto a polyester base. When the film is exposed to ionising radiation, there is polymerisation of the monomers leading to a darkening in the colours. The amount of darkening in the film is proportional to the absorbed dose received by the film. The amount of darkening can be calculated by determining the optical density of the film. There are numerous methods published in the literature on how to determine optical density from the film.

b. A typical calibration curve for the GafChromic EBT3 film is shown in Figure 5.2. It is noted that for this experiment, the raw pixel values of the scanned film are plotted as a function of dose.

For routine dosimetry of radiotherapy beams, the red channel is the most commonly used as it is the highest dynamic range over this dose range. However, the green or blue channel will provide better dose resolution for higher doses typically above 10 Gy.

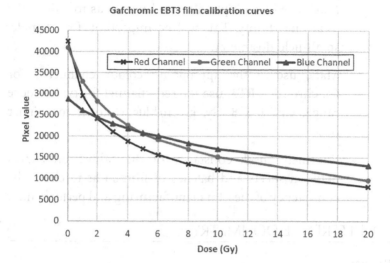

FIGURE 5.2 A typical calibration curve for the GafChromic EBT3 film.

c. Key aspects for a good QA programme for radiochromic film:

- Cut the pieces of film in the same orientation and – otherwise polarisation effects can impact on your optical density reading.

- Minimise markings on the film by using gloves.

- Label each film in the top corner to avoid mix-up of film pieces.

- Warm-up of the desktop scanner by switching on and also performing a number of dummy scans.

- Ensure correct scanner settings within the scanning software as recommended by the film manufacturer, for example, use of transmission mode.

- Ensure that the film orientation is same on the scanner and located within the centre to reduce variations in scanner response.

- Make use of multiple film readings so as to reduce the total uncertainty. This will be important, for example, small-field dosimetry.

- Make use of the appropriate colour channel – for Gafchromic EBT3, use the red channel for typical dose range, e.g., 0.5–8 Gy, green/blue channel for higher doses.

- Routinely check constancy of the calibration curves between sheets and batches of film so as to check the consistency of the film manufacturing and performance.

5.9 REFERENCE DOSIMETRY

Problem:

a. List the main three dosimeters that can be used for reference or absolute dosimetry of radiotherapy beams, i.e. determination of absorbed dose (J/kg) without reference to a known dose.

b. List the most commonly used dosimeters that are used in the clinic for reference dosimetry calibrations of the following radiotherapy:

   i. Megavoltage X-ray beam (dose to water)

   ii. Megavoltage electron beam (dose to water)

   iii. Kilovoltage X-ray beams in energy range from 100 to 300 kVp (air-kerma calibration)

   iv. Kilovoltage X-ray beams with energy range less than 50 kVp (air-kerma calibration)

Solution:

a. Calorimetry, chemical dosimetry (e.g., Fricke dosimetry) and free air ionisation chamber

b. See table below.

Radiation Beam	Detector
i. Megavoltage X-ray beam (dose to water)	Cylindrical ionisation chambers
ii. Megavoltage electron beam (dose to water)	Plane-parallel ionisation chamber
iii. Kilovoltage X-ray beams in energy range from 100 to 300 kVp (air-kerma calibration)	Cylindrical ionisation chamber
iv. Kilovoltage X-ray beams with energy range less than 50 kVp (air-kerma calibration)	Low-energy X-ray plane-parallel ionisation chamber

5.10 PHANTOM MATERIALS

Problem:

a. Define what is meant by a phantom material.

b. Explain why water is the most commonly used phantom material in radiation oncology setting.

c. What are the important physical parameters and radiological properties for a material to be considered a phantom?

d. Explain the purpose of solid block phantoms in the radiotherapy clinic and list their application.

Solution:

a. A phantom is defined as any material which is considered radiologically equivalent to the body tissue of interest. Radiological equivalence is determined by evaluating both the absorption and scattering properties for the particular radiation beam of interest.

b. Water is one of the main components of many soft tissues in the body which means that many tissues have a density value that is close to water. This means that the radiation interactions in the soft tissues are very similar to those in water such as the dose absorption and scattering.

So it is for this reason that water is the standard phantom material for the dosimetry of radiotherapy beams. These radiation beams include megavoltage X-ray and electron beams, kilovoltage X-ray beams, brachytherapy sources and particle beams. The published reference dosimetry CoP such as the IAEA TRS398 is based on using a primary standard for dose in terms of dose to water. In addition, the measurement of relative dosimetry data as required for commissioning, quality assurance and treatment planning purposes in the radiotherapy clinic is also recommended to be performed in water.

Water is readily available to use, has consistent density, it is transparent and liquid which makes it easy when placing a dosimeter inside a water phantom.

c. The three important parameters for a phantom are:

- mass density ρ (kg m^{-3}),

- electron density in terms of number of electrons per gram and

- effective atomic number Z_{eff} which depends on the atomic composition of the mixture of different elements in the material.

  The important properties of a phantom material are:

  - For photon beam dosimetry – the mass energy absorption coefficient, mass energy attenuation coefficient, the mass stopping power and mass scattering power.

  - For electron beam dosimetry – the collisional and radiative stopping powers.

d. While water is considered to be the reference phantom material for the dosimetry of radiotherapy beams, there are particular situations where it is not ideal. These include that water can be difficult to handle and often contained in large water phantoms which are heavy and take time to set up for any dose measurements. Many radiation detectors are not waterproof and so cannot be readily placed within the water phantom. In addition, the water phantom itself provides a planar geometry giving a flat surface which is not always in the shape of a patient.

Solid phantoms are found in radiotherapy clinics and are usually in the form of a plastic made from epoxy-resin, PMMA or polystyrene. These solid phantoms should be evaluated to determine their radiological water equivalence in terms of their absorption and scatter properties as compared to water. Examples of solid phantoms found in the clinic have various trade names such as Solid Water, Plastic Water and Virtual Water. There are also solid block phantoms which are radiologically equivalent to the bone- or the lung-type tissues.

Solid block phantoms are used for dosimetry measurements such as (i) routine dose outputs; (ii) relative dose measurements such as total scatter factors, wedge factors and electron cut-out factors; (iii) as part of check devices which can measure profiles and energy parameters as part of routine quality assurance measurements; and (iv) calibration of *in vivo* dosimeters such as radiochromic film or OSLDs.

Solid block phantoms are also used for independent checks of a radiotherapy plan for a patient. Within the radiotherapy treatment planning system, the patient plan is mapped over onto a block phantom and the doses recalculated without changing any of the plan settings. The patient plan is delivered to the block phantom on the linear accelerator. The doses can be measured inside of the phantom using small volume ionisation chambers or radiochromic film if a 2D dose plane is required. These checks are done at the time of commissioning as well as part of routine patient plan quality assurance.

5.11 IAEA TRS398 DOSE CALIBRATION (1)

Problem:
Write the equation that is defined in the IAEA TRS398 CoP that is used to define the absorbed dose in water under reference conditions for a user beam with a beam quality Q. List the influence quantities which are used to correct the reading of the dosimeter to their reference values.

Solution:
The absorbed dose to water, $D_{w,Q}$, at a reference depth z_{ref} for the beam with quality Q is given by the following equation:

$$D_{w,Q} = M_Q \, N_{D,Q_0} \, k_{Q,Q_0}$$

where

M_Q is the reading of the dosimeter in the beam with quality Q that is corrected for the reference values of the influence quantities (with the exception of beam quality),

N_{D,Q_0} is the calibration factor of the dosimeter in terms of absorbed dose to water for the reference beam quality Q_0 which has been provided by the standards laboratory and

k_{Q,Q_0} is the beam quality correction factor which corrects for the effects of the difference between the reference beam quality Q_0 and the beam quality in the clinic Q.

The key influence quantities that require corrections to be applied to the dosimeter reading are:

• Air temperature, pressure and humidity

• Ion recombination

• Polarity effect

• Electrometer calibration factor

5.12 IAEA TRS398 DOSE CALIBRATION (2)

Problem:
The following set of data is provided to you in order to calculate the dose in water for a 6 MV X-ray beam. We will be using the IAEA TRS 398 CoP (Version 12, 5 June 2006). This can be accessed via this link and will be needed for the dose calculation: http://www-naweb.iaea.org/NAHU/DMRP/codeofpractice.html.

The PDF version can be directly found here: http://www-naweb.iaea.org/NAHU/DMRP/documents/CoP_V12_2006-06-05.pdf.

A dose calibration is being performed for a 6 MV radiotherapy linear accelerator. The local clinical reference conditions are such that the linear accelerator is calibrated in a fixed SSD geometry according to the following conditions:

- 1 Gy = 100 monitor units for an SSD = 100 cm, depth of D_{max} (1.5 cm) and field size of $10 \times 10 \, cm^2$.

- Gantry = 0.0° and collimator = 0.0°.

- Reference field size of $10 \times 10 \, cm^2$ using a combination of the jaws and MLCs as per the vendor requirements for calibration and the treatment planning system.

- Dose rate of 600 monitor units per minute.

The dose calibration is performed with the following equipment:

- PTW 30013 Farmer-type ionisation chamber (SN: 10001)

- PTW UNIDOS Electrometer (SN: 0002)

- PTW MP3 Scanning water tank (SN: W0001)

This Farmer chamber has a calibration coefficient provided by an SSDL based on a dose-to-water calibration using a Co-60 source:
ND, w = 53.74 mGy/nC

(*Note:* This calibration coefficient includes the polarity and recombination corrections for Co-60).

This electrometer has a calibration coefficient $k_{elec} = 0.999$ nC per nominal charge reading. These were both performed at voltages of $V = +400\,V$ and corrected to reference atmospheric conditions of $T_0 = 20.0°C$, $P_0 = 101.3\,kPa$ and 50% humidity.

PDD Data for the $10 \times 10\,cm^2$ Field at SSD = 100 CM

Depth (cm)	Relative Dose
1.5	100.0%
5.0	85.7%
10.0	66.1%
20.0	37.9%

The following readings were collected with the Farmer chamber positioned in the water following the IAEA TRS398 CoP methodology:

- SSD = 100 cm

- Depth of 10 cm to the P_{eff} of the chamber being the centre of the chamber

- Each reading was taken for 100 monitor units

- Nominal voltage of +400 V

Voltage to the Chamber	Reading 1 (nC)	Reading 2 (nC)	Reading 3 (nC)
+400 V	12.41	12.40	12.40
−400 V	−12.40	−12.40	−12.39
+100 V	12.30	12.29	12.30

Air temperature = 19.8°C

Air pressure = 101.0 kPa

Use this provided information to determine the dose per 100 MU under the local calibration conditions.

Solution:
We will proceed through TRS398 CoP in Chapter 6 to determine the dose in water for this 6 MV X-ray beam. The absorbed dose to water, $D_{w,Q}$, at a reference depth z_{ref} for the beam with quality Q is given by the following equation:

$$D_{w,Q} = M_Q \, N_{D,Q_0} \, k_{Q,Q_0} \tag{5.1}$$

It should be noted that you may get slightly different calculated values due to rounding or methods of interpolation. These should only be a small difference.

The nominal beam quality Q is 6 MV while Q_0 is Co-60. The beam quality index is given by the ratio $TPR_{20,10}$. As we have only PDD data provided, we can calculate this by the following relationship in the footnote [25], where

$$TPR_{20,10} = 1.2661 \, PDD_{20,10} - 0.0595$$

As $PDD_{20,10} = (37.9/66.5) = 0.5734$, then $TPR_{20,10} = 0.6664$.

The chamber dose calibration factor using a Co-60 beam is given by $N_{D,w} = 53.74$ mGy/nC or 0.05374 Gy/nC.

The value of k_{Q,Q_0} can be found in Table 6.III of the CoP noting that we have a PTW 30013 ionisation chamber and that $TPR_{20,10} = 0.6664$. A simple interpolation gives $k_{Q,Q_0} = 0.992$ – you may get slightly different value for the third decimal place depending on your interpolation method.

We now calculate M_Q which is the charge reading on the electrometer with all influence quantities taken into account. That is:

$$M_Q = M_{raw} \times k_{TP} \times k_{elec} \times k_{pol} \times k_s \tag{5.2}$$

We will now go through each of these individual factors.

M_{raw} is the mean reading at the nominal voltage (being +400 V) and has a value of 12.403.

k_{TP} is the temperature-pressure correction factor (Section 4.4.3.1) and is given by the following equation:

$$k_{TP} = \frac{273.2+T}{273.2+T_0} \frac{P_0}{P}$$

As listed in the provided data, we have an air temperature $T=19.8°C$ and air pressure $P=101.0$ kPa.

Using the reference values of $T_0=20.0°C$ and $P_0=101.32$ kPa gives us a value for $k_{TP}=1.003$.

The provided calibration certificate has $k_{elec}=0.999$ – Note: that this is valid for the particular settings on the electrometer, e.g., certain range and charge reading mode. Details of this factor are found in Section 4.4.3.2 of the TRS398 CoP.

k_{pol} is the polarity correction factor (see Section 4.4.3.3) and is given by:

$$k_{pol} = \frac{|M_+|+|M_-|}{2M}$$

M_+ is the mean reading at positive polarity (+400 V in this case), M_- is the mean reading for negative polarity being −400 V and M is the mean reading for the usual voltage used which in this case is +400V.

The mean readings are calculated to be $M_+=12.403, M_-=-12.397$ and $M=12.403$.

This gives a value of $k_{pol}=1.000$ (to 3 decimal places).

k_s is the recombination correction factor (see Section 4.4.3.4) and is calculated by the two-voltage method using the following equation:

$$k_s = a_0 + a_1\left(\frac{M_1}{M_2}\right) + a_2\left(\frac{M_1}{M_2}\right)^2$$

M_1 is the mean reading at the usual polarity being +400 V in this case and M_2 is the mean reading for the reduced polarity being +100 V. The mean readings are calculated to be $M_1 = 12.403$ and $M_2 = 12.297$.

The values for a_0, a_1 and a_2 are taken from Table 4.VII noting that a linear accelerator is a pulsed beam. The ratio of the two voltages is 4.0 and so we use values of $a_0 = 1.022$, $a_1 = -0.363$ and $a_2 = 0.341$. This gives a value of $k_{pol} = 1.003$ to 3 decimal places.

Using Equation 5.2, the final corrected meter reading is given by:

$$M_Q = 12.403 \times 1.003 \times 0.999 \times 1.000 \times 1.003 = 12.465 \text{ nC} / 100 \text{ MU}$$

Referring back to Equation 5.1, we can then calculate the dose in water at the measurement depth of 10 cm and is given by:

$$D_w = 12.465 \text{ } nC / 100 \text{ MU} \times 0.05374 \text{ Gy/nC} \times 0.992$$

$$= 0.6645 \text{ Gy} / 100 \text{ MU}$$

However, we need to correct to the dose at the reference point used in the clinic. In this case, this is at the depth of $D_{max} = 1.5$ cm. We do this by correcting with the value of *PDD* as follows:

Dose at 1.5 cm depth=dose at 10 cm depth/*PDD* (10 cm)= 0.6645/0.661 = 1.005 Gy per 100 MU.

According to the AAPM TG-142 report, the absolute or reference dose for photon beams should be within ±1%. In this case, we are within 0.5% and so meet that criteria. However, some radiotherapy departments may choose to adjust the linear accelerator output to give exactly 1.000 Gy per 100 MU at the time of the annual dose calibration.

5.13 BEAM QUALITY CORRECTION FACTOR

Problem:
Explain the purpose of the beam quality correction factor k_{Q,Q_0} and how it is calculated.

Solution:
The beam quality correction factor k_{Q,Q_0} is a correction for the ionisation chamber in terms of its response in a reference beam quality Q_0 and the beam quality in the radiation oncology clinic Q. For many standards labs, the reference beam quality Q_0 is Co-60. However, it is noted that some standards labs will offer calibrations in terms of megavoltage X-ray and electron beam qualities.

The beam quality correction factor k_{Q,Q_0} is calculated by the following equation:

$$k_{Q,Q_0} = \left(S_{air}^w\right)_{Q_0}^Q \left(W_{air}\right)_{Q_0}^Q \frac{P_Q}{P_{Q_0}}$$

where

Q is the beam quality of the particular beam in the clinic,

Q_0 is the reference beam quality which is usually Co-60,

$\left(S_{air}^w\right)_{Q_0}^Q$ is the mean restricted collisional stopping power ratio of water to air averaged over the spectra for the two beam qualities Q and Q_0,

W_{air} is the average energy to create an ion pair in air and has a value of 33.97 eV for both beam qualities and

P_Q and P_{Q_0} are the ionisation chamber perturbation correction factors for the two beam qualities Q and Q_0.

5.14 IONISATION CHAMBER PERTURBATION CORRECTION FACTOR

Problem:
List and explain the factors that are used to determine the ionisation chamber perturbation correction factor P_Q.

Solution:
The ionisation chamber perturbation correction factor P_Q is defined by the following equation:

$$P_Q = (P_{dis} P_{wall} P_{cel} P_{cav})_Q$$

where

P_{dis} corrects for the displacement of the effective point of measurement of the ionisation chamber upstream from the central axis of a cylindrical chamber,

P_{wall} corrects for the difference between the chamber wall material and water,

P_{cel} corrects for the material used for the central electrode of the ionisation chamber and

P_{cavl} corrects for influence and differences in perturbation of the electron fluence due to the differences between water and air.

5.15 DETECTOR ARRAYS

Problem:
The features of diodes listed in question 5.7 lend themselves to a combination of multiple diodes into a detector array. Such an array can have hundreds or thousands of detectors. They are commonly used for treatment plan verification for individual patients using gamma index analysis. Describe what is gamma index and list parameters that need to be considered when deciding on the gamma criterion for the determination of pass and fail rates.

Solution:
The gamma index (γ) combines dose difference and distance to agreement (DTA) to calculate a dimensionless metric for each point in the evaluated distribution. It compares dose differences at specific points and determines the distance to the nearest point having the same dose value. The dose difference criterion is important in low-dose gradient regions, while yields valuable information in regions having a high-dose gradient. If $\gamma \leq 1$, the dose measured at the point passes the criteria of the dose difference and DTA. If $\gamma \geq 1$, the dose measured at the point is out of the acceptable tolerance and fails the criteria of the dose difference and DTA. Dose difference and DTA criteria can be selected and often values of 3%/3mm are used.

The reference dose distribution is generally taken as the 'gold standard', e.g., it could be the dose distribution that has been measured. In theory the distribution could be a single point (e.g. ionisation chamber measurement), 1D (e.g., a line profile), 2D (e.g., film measurement) or 3D (e.g., gel dosimetry, Monte-Carlo simulation). The evaluated dose distribution is what is being compared. In most cases this will be the predicted TPS dose distribution that is being checked for accuracy in modelling the delivered dose.

The following parameters should be considered when deciding the gamma criterion:

- Acceptance criteria (e.g., 3 mm/3% or 1 mm/3% for stereotactic procedures)

- Spatial resolution of the detector and calculated dose distribution (dose grid)

- Absolute versus relative gamma (is there a scaling of the detector values?)

- Local and global gamma (is the dose criterion in % of maximum or local dose?)

- Thresholds (do we ignore dose values below a certain threshold – e.g., 10%?)

5.16 SMALL-FIELD DOSIMETRY

Problem:
Small fields in radiotherapy are used in various treatment techniques such as Intensity Modulated Radiotherapy, Volumetric Modulated Arc Therapy and Stereotactic Radiosurgery. List the three physical conditions by which photon beams are considered to be small fields.

Solution:
Three physical conditions by which photon beams are considered to be small fields:

- There is a loss of lateral charged particle equilibrium on the beam central axis. This occurs if the radius of the beam is smaller than the range of the majority of secondary electrons of the beam.

- There is partial occlusion of the photon source. This is where the finite size of the primary photon beam source (focal spot) can be partially shielded by the beam collimation from both MLCs and collimator jaws. This typically occurs at the very small field sizes.

- The size of the detector is similar or larger than the size of the X-ray beam. This results in the process known as volume averaging whereby the detector is measuring dose in a large dose gradient resulting in a reduction in signal.

CHAPTER 6

Special Procedures and Advanced Techniques in Radiotherapy

6.1 TOTAL BODY IRRADIATION

Problem:

a. Explain briefly the clinical purpose of total body irradiation (TBI) and how you would deliver this treatment to the patient.

b. Describe how you would verify the dose delivered to the patient for each treatment session.

DOI: 10.1201/9780429159466-6

93

Solution:

a. TBI involves delivering a whole-body radiation dose in order to eradicate off all the bone marrow cells. This irradiation is usually delivered just prior to a bone marrow transplant. There are several dose regimes provided depending on the particular condition and other therapies including chemotherapy. The prescribed dose is typically in the range from 2 Gy in 1 fraction up to 12 Gy in 6 fractions delivered over 3 days.

These treatments are typically delivered with lower dose rates of 7–8 cGy per minute – this is done to reduce toxicity to the patient and maximise tumour cell kill for the bone marrow.

There are several technical challenges in delivering the X-ray beams to the whole body:

- Ensuring dose coverage over the whole body.

- Delivering a lower radiation dose rate to the patient.

  This is typically achieved by having the patient positioned at a large SSD of several metres away from the isocentre and setting the collimator angle to be 45°. This means that the projection of the X-ray field is maximised to cover the whole patient. This also means that due to the inverse square law, the dose rate is reduced. The linac may need to be operated in a lower dose rate, e.g., 100 or 200 MU/minute in order to obtain the required dose rate.

  There are various possible patient geometries that can be used for the TBI treatment. The most commonly used would be having the patient lying on a specially made couch with the beams coming in from both sides of the patient. As the patient treatment session can be typically up to 1 hour in total, this arrangement provides a comfortable set-up. There are other possible set-ups including having the patient standing to face the beam or having a moving treatment couch at an extended SSD.

b. Verification of the doses delivered to the patient can be done by in *vivo* dosimetry measurements. There are several detectors that would be suitable for this including radiochromic film, diodes, MOSFETs, TLDs or OSLDs. Each of these has its respective advantages such as having a permanent dose record and instantaneous dose read-outs during the treatment. However, the performance of the dosimeter must be checked while being used in low dose-rate conditions.

6.2 INTENSITY-MODULATED RADIOTHERAPY

Problem:

a. Describe two modes of MLC for intensity-modulated radiotherapy (IMRT) delivery.

b. Is the leakage radiation higher or lower with IMRT compared to conventional radiotherapy? Give reasons for your answer.

Solution:

a. Two modes of MLC for IMRT delivery:

- Segmented (step and shoot or static) – the intensity-modulated fields are delivered with a sequence of small segments of subfields, each subfield with a uniform intensity. The beam is only turned on when the MLC leaves are stationary in each of the prescribed subfields positions. There is no MLC motion when the beam is turned on.

- Dynamic (sliding window) – intensity-modulated fields are delivered in a dynamic fashion with the leaves of the MLC moving during the irradiation of the patient. For a fixed gantry position, the opening formed by each pair of opposing MLC leaves is swept across the target volume under computer control with the radiation beam turned on to produce the desired fluence map.

b. IMRT delivers much higher MU per field compared to conventional radiotherapy in achieving the same prescribed dose to the target. The total number of MU for IMRT is typically 2–5 times that for conventional treatments. With IMRT, there is also increased beam-on time to deliver modulated fields. As a result, the leakage radiation from the linac head is much higher in IMRT compared to conventional radiotherapy.

6.3 TOTAL SKIN ELECTRONS

Problem:
Describe the use of total skin electron treatments and explain how these may be delivered to a patient. Provide details of the suggested beam properties and geometry.

Solution:
Total skin electron treatments are used to irradiate all the skin of a patient for conditions such as mycosis fungoides. By using a low-energy electron beam, this treatment spares the rest of the body including the sensitive organs. There are two possible ways of irradiating the patient:

- The patient sits or stands on a continuously rotating platform.

- The use of multiple large fields.

The beam properties are as follows:

- Electron beam energy of 6–10 MeV.

- Nominal SSD of 300–500 cm.

- Electron field size typically of $80 \times 200 \, \text{cm}^2$ to ensure the field covers the whole patient.

- Dose uniformity at depth of D_{max} to be within 5% across at least 80% of the field area.

6.4 IMAGE-GUIDED RADIOTHERAPY

Problem:

a. Describe what is meant by the term 'Image-Guided Radiotherapy' (IGRT) and its purpose in clinical use?

b. List the main forms of currently available IGRT methodologies available on linear accelerators along with key advantages.

Solution:

a. Image-guided radiation therapy is the process by where radiographic imaging is taken before and during a session of radiotherapy. This is done to ensure that the patient is localised to be in the same position as at the time of their radiotherapy simulation based on their CT imaging and the radiotherapy treatment plan.

The purpose of IGRT is several-fold. The primary goal of IGRT is to increase the local control of the treatment and reduce any short- and long-term side effects. This imaging is taken to ensure that the radiation beams are directed to the tumour volume. This is needed due to day-to-day variation of the patient positioning on the treatment couch as well as interaction and intra-fraction organ motion.

The imaging will make use of the patient anatomy by either boney landmarks or matching up to well-defined organs that are readily visible in the imaging scans taken. Once the imaging is taken, the treatment couch position is shifted until the patient is within the correct location to within certain tolerances, e.g., 2 or 5 mm. This means that when the radiation beams are switched on, the patient is located in the correct position. This is particularly important for IMRT/VMAT treatments where there is large dose modulation and tight treatment margins. It is also important for treatments which involved large prescribed doses in only a few or a single treatment fraction.

b. The most commonly used methods for IGRT are:

- Matching megavoltage (MV) images with the DRRs calculated from the planning CT scans.

- Matching planar kilovoltage (kV) radiographs with digital reconstructed radiographs (DRRs) from the planning CT scans of the patient.

- Matching cone beam computed tomography (CBCT) data set with the planning computed tomography (CT) data set from planning.

The advantage of the MV image matching is that it uses the actual X-ray beam used for treatment with the electronic portal imager device (EPID). This allows direct transmission imaging through the patient and possibly during the actual treatment delivery. The major disadvantage of this imaging modality is that the image contrast is inferior compared to kV imaging.

The kV image matching is performed using an on-board imager which is attached to the linear accelerator. But delivering kV X-ray beams, very good image quality is achieved with high image contrast between the different tissue types. Typically, imaging matching is performed with two X-ray beams at 90° to each other, e.g., first an A/P direction beam and then one lateral beam L/R. This form of imaging matching is very fast and so would be suitable for busy radiotherapy departments.

CBCT is achieved by using the on-board imager (OBI) unit on the linac with multiple planar exposures which are then reconstructed into a 3D CBCT image data set. The advantage of CBCT is that is provides 3D image information and theoretically would be expected to be more accurate compared to kV planar imaging. The disadvantages of CBCT imaging include the fact that these take longer to image and reconstruct and that the image quality for the CBCT is not as good as that from a CT scanner.

6.5 STEREOTACTIC ABLATIVE BODY RADIOTHERAPY

Problem:
Describe the key characteristics of a Stereotactic Ablative Body Radiotherapy (SABR) treatment. Note that SABR is also known as Stereotactic Body Radiotherapy. List which treatment sites are typically treated by SABR.

Solution:
The following are specific requirements and characteristics for SABR treatments:

- SABR is a form of radiotherapy that is hypo-fractionated. By this we mean that we deliver a high-radiation dose in small number of treatment fractions (typically 1–5).

- Dose delivery to the PTV with millimetre accuracy.

- Appropriate patient immobilisation to reduce intra-fraction and inter-fraction motion for the patient.

- Application of motion management techniques to reduce organ motion.

- Reproducible patient positions at the time of simulation and during the treatment.

- Treatment planning process which maximises the dose to the PTV takes into account the ITV and minimises doses to the organs at risk as well as healthy tissues.

- Daily imaging for each treatment fraction typically using CBCT.
    Treatment sites which are treated with SABR include lung, liver, kidney, pancreas, spine and prostate.

6.6 PROTON THERAPY

Problem:

a. Explain how the Bragg peak is formed in matter.

b. Draw a diagram comparing the depth dose curves of conventional high-energy X-ray therapy with proton therapy.

Solution:

a. When a fast-charged particle travels through matter, it ionises atoms of the material and deposits energy along its path. A peak occurs because the interaction cross section increases as the charged particle's energy decreases. For protons (and other charged particles), a sharp and narrow peak occurs immediately before the particles come to rest. This is called the Bragg peak.

b. See Figure 6.1.

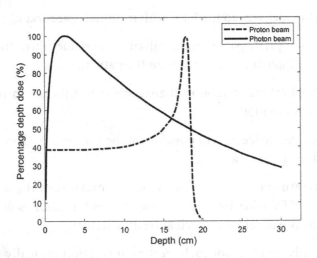

FIGURE 6.1 Comparison of depth dose curves of conventional high-energy X-ray therapy with proton therapy.

6.7 SURFACE-GUIDED RADIOTHERAPY

Problem:
Describe the features of surface-guided radiotherapy (SGRT) equipment and how it can be used in clinical practice?

Solution:
SGRT makes use of optical imaging systems to allow imaging of the patient to assist with set-up and monitoring during the treatment. These optical imaging systems, which emit either visible light or infrared, are usually mounted on the wall or ceiling facing towards the gantry. This reconstruction generates a 3D surface map of the patient in real time using stereoscopic cameras. These systems have the advantage of performing patient imaging without the need to deliver additional radiation doses. Some imaging systems make use of optical or IR markers which may be fixed on the patient or onto a fixed frame attached to the couch. These markers can be used as reference points during the imaging.

The patient surface imaging can be used for the initial patient set-up and positioning on the linear accelerator couch. The 3D surface image collected is compared to a similar image set which would be collected during the simulation session. The surface imaging can also be used during the treatment to take into account any infra-fractional motion of the patient.

SGRT is ideally suited for a number of different radiotherapy techniques such as breast and lung treatments. SGRT is also used in stereotactic radiosurgery as a key aspect of patient monitoring of the head in conjunction with the X-ray imaging which is an important key aspect of accurate patient treatment due to the very small tumour volume sizes and high-dose gradients.

6.8 BREATHING CONTROL IN RADIOTHERAPY

Problem:

a. Describe the reasons why breathing control needs to be considered.

b. List the different methods by which the respiratory motion can be quantified in radiotherapy.

c. List the main radiotherapy treatment sites which are most affected by respiratory motion.

Solution:

a. The process of breathing by inspiration and expiration causes breathing-related motion in the patient. This means that both the tumour volumes and organs at risk move during the patient's treatment. As such, this motion needs to be taken into account either by consideration of the margins used or by breathing control. There are a number of specific volume definitions that can be used to evaluate the lung function of the patient. A breathing trace can be taken in order to evaluate this lung function and the changes in volumes.

b. There are a number of methods and technologies that are used in the clinic to help quantify the amount of respiratory motion. Examples are as follows:

- Spirometry

- CT imaging

- Surface-guided radiotherapy (SGRT) techniques

- Deep inspirational breath hold (DIBH)

- MRI

c. Respiratory motion is mostly evident in the region near the diaphragm and the chest. This means that breathing control can impact on treatment sites including the lung, liver, pancreas, chest and breast.

CHAPTER 7

External Beam Commissioning and Quality Assurance

7.1 ACCEPTANCE TESTING, COMMISSIONING AND QUALITY ASSURANCE

Problem:
Explain what is meant by the terms acceptance testing, commissioning and quality assurance with regards to a radiotherapy linear accelerator.

Solution:
Acceptance testing
Acceptance testing is required to ensure that the linear accelerator meets the technical specifications specified by the manufacturer as any additional requirements from the radiation oncology department. This process involves ensuring that the linear accelerator meets all mechanical, electrical, safety, radiation safety and functionality checks. The performance of the linear accelerator must be meeting the specifications as outlined by the manufacturer in

DOI: 10.1201/9780429159466-7

their acceptance testing documentation. Another important part of acceptance testing is to ensure that all of the purchased hardware and software components of the linear accelerator are delivered to the local department.

The acceptance tests can be split into divided into three groups: safety checks, mechanical checks and dosimetry measurements. These tests are performed by the local physicist with the representative from the company that supplied the linear accelerator.

Commissioning

Commissioning of the linear accelerator can only begin once the acceptance testing has been completed and the linear accelerator has been officially handed over to the radiation oncology department. The term commissioning refers to the process of which the performance of the linear accelerator over the whole range of possible operation is completed. This involves collection of data, preparing protocols and instruction as well as giving training so that the linear accelerator can be safely released for clinical service. After this time, the linear accelerator is considered to be safe to treat patients.

Clinical commissioning involves the collection of dosimetry data which is used for beam characterisation and quality assurance testing. This is required for both treatment planning and reference dosimetry. Relative dosimetry measurements are also performed for beam characterisation and use in treatment planning data. Reference dosimetry is performed for each of the X-ray and electron beams that are made available on the linac. QA parameters are measured and established. This allows benchmark parameters to be set which will be used to monitor the performance of the linear accelerator by routine quality assurance tests.

Quality Assurance

According to the AAPM TG142 report, the goal of a QA programme for a linear accelerator is to assure that the machine characteristics do not deviate significantly from their baseline values acquired at the time of acceptance and commissioning.

Within this QA programme, each of machine characteristics will have particular tolerances in which the linac can operate. If any of the characteristics are out of this tolerance, remedial action needs to occur before it can be used again for clinical use.

7.2 COMMISSIONING OF A CT SIMULATOR

Problem:
List the key tests you would need to perform in commissioning a CT scanner to be used in a radiotherapy centre. Your answers should include safety, mechanical, optical and imaging tests.

Solution:
  Safety checks:

- Radiation survey of the CT room
- Check that all radiation warning lights are working correctly
- Check mechanical movements of the couch including any emergency stops
- Dose measurements using CT dose index (CTDI) or other suitable dose parameters

Mechanical and optical tests:

- Ensure that the CT scanner was delivered with a flat couch-top that matches the couch-tops on the linear accelerators
- Check the accuracy of the movements of the couch – vertically and horizontally into the gantry of the scanner
- The couch-top should have an indexing system for which patient immobilisation devices are attached which should be consistent with the linear accelerator
- Check the scan plane of the CT scanner which is the 2D plane inside the gantry of the scanner

- The isocentre of the CT scanner should be positioned in the scan plane

- The internal lasers inside of the CT scanner gantry should be coincident with the isocentre

- The external lasers which typically consist of ceiling and side wall lasers should be orthogonal with each other, aligned to the isocentre and aligned to the scan plane

- The external lasers should also be aligned with the internal lasers

- Check 4DCT scanning functionality with moving phantoms

Imaging tests:

- Development of scanning protocols to be used for different treatment sites, e.g., kVp, mA, slice width, image reconstruction options including any metal artefact reduction algorithms

- Collection of a data set of CT number to relative electron density information to be input into the radiotherapy planning system

- Test the different software functions like windowing level

- Test of spatial resolution, image quality, high-contrast and low-contrast resolution, slide width as well as reconstruction accuracy using a suitable phantom, e.g., CATPHAN phantom

- Test connectivity and accuracy of images by using image export to the radiotherapy planning system

7.3 ABSORBED DOSE CALIBRATION

Problem:
As a physicist in a radiotherapy department, you are required to perform an absolute dose calibration of a 6 MeV electron beam using ionisation chamber measurement.

a. What type of ionisation chamber will you be using and why?

b. Name and describe three correction factors to be applied to the readings from the electrometer.

c. You have decided to calibrate the beam based on the IAEA TRS398 formalism using an ionisation chamber. The ionisation chamber is traceable to the secondary standard laboratory which is calibrated with a Co-60 unit. 100 MU are delivered to a reference point in water under a reference condition. How is the reference depth for positioning the ionisation chamber determined?

d. Suppose the corrected reading from the electrometer is 11.60 nC, and calculate the absolute dose at the reference point. (Given $N_{D,w} = 8.959$ cGy/nC, $k_{Q,Co-60} = 0.9575$, $PDD_{1.3cm} = 100\%$ and 1 MU = 1 cGy under reference conditions.)

Solution:

a. The IAEA TRS398 CoP recommends the use of a parallel plate ionisation chamber for calibrations of electron beams. Electron beams can have a steep depth dose gradient along the central axis of the beam. Parallel plate ionisation chamber is constructed of better localised volume (compared to thimble chamber) that still contains enough air to collect enough charge at clinically used dose rates.

b. Temperature and pressure correction factor – Temperature (the higher the less molecules in chamber). Pressure (the higher the more molecules in chamber).

Polarity correction factor – Under identical irradiation conditions, the use of polarising potentials of opposite polarity in an ionisation chamber may yield different readings, a phenomenon that is referred to as the polarity effect.

Recombination factor – Ionisation chambers are commonly used in the near-saturation region or even in the saturation region where all charges produced by radiation are collected. When the chamber is used below saturation, some of the charges produced by radiation recombine and are lost to the dosimetric signal.

c. Reference depth for positioning the ionisation chamber, $D_{ref} = 0.6 R_{50} - 0.1$ g/cm^2, with R_{50} given in g/cm^2.

d. Dose to water = [Corrected reading $\times N_{D, w} \times k_{Q, Co\text{-}60}$]/ $PDD_{1.3cm} = 99.51$ cGy.

7.4 MONTHLY QA PROGRAMME ON A LINEAR ACCELERATOR

Problem:
List the key tests that you would perform on a monthly basis on a linear accelerator. This machine is typically used for a range of radiotherapy treatments using 3D conformal, IMRT and VMAT plans.

Solution:
The following list of tests are some of those that should be considered as part of the monthly QA programme.

Types of Check	List of Tests
Safety checks	• All safety interlocks • Audio-visual system • Radiation warning lights • Check of the last-person out system • Laser guard system • Neutron door – for linear accelerators with X-ray beam energies > 10 MV
Mechanical checks	• Gantry, collimator and couch angles • Couch absolute positions and movements • Collimator jaw position accuracy for symmetric and asymmetric fields • MLC position accuracy checks • MLC speed tests • Isocentre checks • Fixed wedge interlocks

(Continued)

Types of Check	List of Tests
Optical checks	• Cross-hairs • Lasers • Optical distance indicators • Light field sizes • Radiation/light field coincidence
Dosimetry checks	• Dose outputs for photons and electrons • Beam flatness and symmetry for photons and electrons • Beam energy check for photons and electrons • Wedge factors for both fixed and dynamic wedges • Dynamic MLC dose outputs for some reference fields • Respiratory gating dose output checks
Imaging	• Planar kV imaging tests • CBCT imaging tests • Planar MV (EPID) imaging tests
Respiratory gating	• Beam control with amplitude and phase modes • Gating interlocks • Check of in-room monitoring systems

The AAPM Task Group 142 report entitled "Quality assurance of medical accelerators" provides guidance on a quality assurance programme for linear accelerators. The report can be found at https://doi.org/10.1118/1.3190392.

7.5 LINEAR ACCELERATOR DOSE OUTPUT CHECKS

Problem:
The AAPM TG142 report provides recommendations on the quality assurance of linear accelerators. This document provides a list of suggested tests that need to be done, the frequency at which the tests should performed and the tolerance values for which remedial action is required such as dose output adjustment. List the recommendations for the checking of the dose outputs on a multimodality linear accelerator which generates both megavoltage X-ray and electron beams. You should indicate the particular equipment that you would use for particular tests.

Solution:
Daily dose constancy check:
    The dose output should be checked on a daily basis to ensure that there are no significant dose calibration errors or problems

with the linear accelerator. This check can be done with an electronic check device which provides a nominal dose reading. The simplest devices have a single detector such as an ionisation chamber or diode for checking the dose output. There are also more sophisticated devices which consist of an array of detectors which also allows for a quick testing of beam flatness, symmetry and/or energy. These devices must be robust, simple to use and provide an immediate dose output value.

Dose tolerances = ±3%

Monthly dose output check:

A monthly dose output check is performed using an ionisation chamber located positioned within a solid phantom. A simple water phantom may also be used. The radiation doses are calculated in Gy per monitor unit with application of temperature and pressure correction factors as well as a block calibration factor. This block calibration factor is determined by an inter-comparison of charge readings at the time of the annual dose calibrations. X-ray beam dose output checks would be usually performed using a Farmer-type ionisation chamber typically at a depth of 5 or 10 cm in the solid phantom. The electron beam check would be performed using a plane-parallel plate ionisation chamber at a depth close to the corresponding D_{max} for that particular beam.

Dose tolerances = ±2%

Annual dose calibration

This is a complete dose calibration using one of the international dosimetry codes of practice such as the IAEA TRS398 CoP or AAPM TG51 protocol, for example. Dose measurements are performed in a water phantom under the reference conditions specified by the code of practice. These measurements are done with an ionisation chamber that has a suitable dose-to-water calibration factor ($N_{D,w}$) provided by one of the standards dosimetry labs. At this time, measurements are performed to set up the daily and monthly dose output checks. This would involve the transfer of the dose values over to these secondary pieces of equipment.

Dose tolerances = ±1% (absolute dose)

7.6 BEAM FLATNESS

Problem:

a. The beam profile in Figure 7.1 is acquired at 10 cm depth, with 100 cm SSD. Estimate the field size for the beam profile.

b. How is beam flatness defined?

c. Estimate the beam flatness for this beam profile.

d. What contributes to the dose at point A on the diagram?

e. What is the recommended frequency for the linear accelerator beam flatness QA and the recommended tolerance limit?

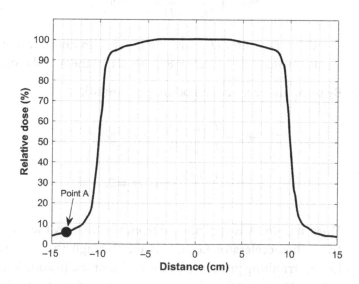

FIGURE 7.1   Profiles of 6 MV X-ray at 10 cm depth.

Solution:

a. Field size is defined as the lateral distance between the 50% isodose lines at 100 cm source-axis distance. From the figure, the lateral distance between the 50% isodose lines for the profile is from −10 to 10 cm. Hence, the field size is 20 cm.

b. There are a number of different methods by which the flatness can be quantified as found in standards reports of QA documents. Two of the most common methods are:

- Flatness can be calculated (D_{max}/D_{min}), where D_{max} and D_{min} are the maximum and minimum dose values in the central 80% region of the lateral profile.

- Another definition of flatness is given by the equation being

$$\%F = \left(\frac{D_{max} - D_{min}}{D_{max} + D_{min}} \right) \times 100$$

where D_{max} and D_{min} are the maximum and minimum dose values in the central 80% region of the lateral profile.

c. By using the second method in question (d),

$$Flatness = \frac{D_{max} - D_{min}}{D_{max} + D_{min}} \times 100\%$$

$$= \frac{100 - 95}{100 + 95} \times 100\% = 2.56\%$$

d. Dose at point A is contributed from the linac by the scatter from the accelerator head, leakage through the jaws and bremsstrahlung produced in the collimators. In addition to this scatter from the phantom/patient contributes to dose at point A, a contribution that increases with depth and field size.

e. In terms of tolerances, it is important to compare (i) the absolute flatness and (ii) the constancy in the flatness. The absolute flatness is measured at the time of commissioning of the X-ray beams and the linear accelerator will also have their specific tolerance. However, for each of the X-ray beam energies, one should be able to achieve a flatness, in both cross-plane and in-plane, of better than 3% for the largest field size of $40 \times 40 \, cm^2$. The regular quality assurance programme of the linear accelerator will include checks of the beam flatness and comparison to the profiles taken at the time of commissioning. For these regular QA checks, the tolerance in the flatness constancy is 1%.

7.7 BEAM SYMMETRY

Problem:
The beam symmetry is an important parameter for the megavoltage X-ray beams. Define the term beam symmetry, describe how the symmetry can be calculated from the lateral beam profiles on a linear accelerator and discuss suitable tolerance values.

Solution:
The symmetry of an X-ray beam is a measure of the consistency of dose on each side of the lateral beam profile and the difference between the maximum and minimum doses within a specified portion of the lateral profile.

There are a number of different methods by which symmetry can be determined. Two of the most common are:

- Symmetry can be calculated as the maximum percentage dose difference (%S) in two points that are equidistant from the beam central axis typically within the central 80% region of the lateral profile.

- Another definition of symmetry is given by the AAPM TG45 report being

$$\%S = \left(\frac{\text{Area}_L - \text{Area}_R}{\text{Area}_L + \text{Area}_R} \right) \times 100$$

where Area_L and Area_R are the total areas on the left- and right-hand sides, respectively, of the 80% width region of the lateral profile.

In a similar manner to beam flatness, there will be tolerances for both the absolute symmetry and the constancy in the symmetry. The absolute symmetry will be determined specified by the manufacturer but should generally be within 2%. The symmetry constancy should be such that the symmetry does not vary by more than 1% as compared to the commissioning or baseline data.

7.8 BEAM FLATNESS AND SYMMETRY MEASUREMENT

Problem:
List the key points in the measurement of beam flatness and symmetry for megavoltage X-ray beams.

Solution:
The beam profile measurements should be performed in a 3D scanning water tank using a suitable ionisation chamber such as the ones typically used for scanning measurements. Two examples of these are the IBA CC13 and the PTW Semiflex 3D detectors. These ionisation chambers provide a good balance between high spatial resolution and sufficient volume to provide a good charge measurement. There will be similar detectors from other manufacturers.

For more routine measurements, such as those performed on an annual basis, 2D dosimetry devices consisting of an array of ionisation chambers or diodes can also be used. It is preferable though that these are taken once water tank scans are completed and can then be used for constancy checks. These 2D arrays are an efficient manner for measuring beam profiles on a monthly basis or as needed if the linear accelerator is being repaired.

For each X-ray beam energy, cross-plane and in-plane beam profile measurements should be performed for a range of depths and field sizes. Typically, during commissioning, a greater number of depths and field sizes maybe be used and will also be based on the recommendations of the linear accelerator manufacturer. For a typical linear accelerator, the minimum data set would be for depths of d_{max} and 10 cm and field sizes of 10×10 cm^2 and 40×40 cm^2 (or the maximum).

Alternative detectors methods for measuring these beam profiles can include the use of electronic portal imagers (EPIs) and radiochromic film. However, these require careful use with appropriate calibrations of the EPI unit or the film along with collection of benchmark profiles taken after water tank scans have been taken.

7.9 HVL FOR KV BEAM

Problem:
For therapeutic kilovoltage X-ray beams, the half-value layer (HVL) is the main parameter that is used to describe the beam quality.

a. Define what is meant by the HVL.

b. What materials are typically used for determining the HVL?

c. The following table provides the nominal air-kerma readings (R) that were collected when performing the HVL measurements on an X-ray beam with a peak potential of 100 kVp. Measurements were performed under narrow beam geometry conditions using a Farmer-type ionisation chamber which had a variation of less than 3% for the air-kerma calibration factors (N_K) over the energy range 70–300 kVp. From these readings, determine the first and second HVLs for this particular beam.

Al filtration thickness (mm)	Air-kerma (R)		
	Reading 1	Reading 2	Reading 3
0.0	310.7	310.4	310.3
0.5	277.1	277.2	277.5
1.0	250.4	250.1	249.6
1.5	227.8	227.4	227.5
2.0	208.5	208.3	208.9
2.5	191.9	191.9	192.1
3.0	177.1	177.6	177.4
3.5	164.4	164.6	164.6
4.0	153.4	153.2	153.4
4.5	142.9	143.2	142.8
5.0	133.8	134.3	134.2
5.5	125.6	125.6	125.7
6.0	116.8	116.4	116.3
6.4	110.9	111.2	110.8
7.0	101.8	102.0	101.9
7.3	98.7	98.7	98.5
7.5	93.8	93.7	93.8
8.0	85.2	85.1	85.1
8.5	78.9	78.9	79.0
9.0	73.1	73.1	73.1
9.5	67.7	67.8	67.7
10.0	62.8	62.7	62.8

d. Explain why the second HVL has a greater thickness compared to the first HVL.

Solution:

a. The HVL is defined as that thickness of a material that attenuates the X-ray beam intensity to 50% of its original value.

b. Typically, aluminium or copper depending on the X-ray beam energy. The metal foils should be of high purity typically better than 99.9%.

c. Relative readings:

Al Filtration Thickness (mm)	Relative Dose Readings
0.0	1.000
0.5	0.893
1.0	0.805
1.5	0.733
2.0	0.672
2.5	0.618
3.0	0.571
3.5	0.530
4.0	0.494
4.5	0.461
5.0	0.432
5.5	0.405
6.0	0.375
6.4	0.357
7.0	0.328
7.3	0.318
7.5	0.302
8.0	0.274
8.5	0.254
9.0	0.236
9.5	0.218
10.0	0.202

The first HVL has a value of 3.95 mm of Al, while the second HVL has a value of 8.65 mm. Note that the calculated HVLs will vary slightly by up to 0.05 mm according to the method of interpolation and the respective curve fit.

d. The second HVL is greater due to the X-ray beam hardening as the beam passes through the Al attenuators. This is due to more preferential attenuation of the lower energy photons for the incident X-ray beam spectra.

7.10 COMMISSIONING OF PATIENT IMMOBILISATION DEVICES

Problem:

a. What is the purpose of patient immobilisation devices?

b. What are the steps you would take to commission a new patient immobilisation device into clinical use?

Solution:

a. A patient immobilisation device is made of low-density material, such as carbon fibre, which is fixated onto the linear accelerator treatment couch. The patient is then positioned onto or into this device. The purpose of the immobilisation is to hold a patient in place during their radiotherapy treatment so as to reduce patient movement and keep them fixed in the correct location. This is done to ensure the accuracy and reproducibility of the patient's set-up from simulation, treatment planning and for each treatment session.

b. The key steps in commissioning a new immobilisation device include the following:

- Review the literature and published recommendations such as the AAPM TG 176 report 'Dosimetric effects caused by couch-tops and immobilization devices'. This is available for free download.

- Review the equipment when it arrives in the department for any damage or cracks.

- Review the safety instructions and manual provided with the device including any limitations.

- Check for any high-density components such as components made from metals which will cause artefacts in kV imaging and dosimetric changes if placed within the radiation beam.

- Check that the indexing on the devices matches the indexing on the linear accelerator couch – this ensures that the device is located in the correct position.

- Perform a CT scan of all the components provided with the devices to check the consistency of the material used, e.g., low-density foam, plastic and carbon fibre.

- Import the CT images into the radiotherapy treatment planning system to check the Hounsfield Units of the entire device.

- Perform dose calculations within the device for the X-ray beams and electron beams for which it will be used.

- Perform transmission dose measurements for the radiation beams for which it will be used – this should be done for a range of field sizes and gantry angles.

- Check the surface doses and build-up effects caused by the immobilisation device and its effect on patient dosimetry.

- Determine how the impact of the immobilisation device will be considered in patient dose calculations – will it be included in the patient contour or as a correction factor?

- Review whether the radiotherapy planning system has a library of immobilisation devices to be used for patient calculations.

- Evaluate the performance of the new devices for the first patients to ensure that they meet clinical requirements.

- Develop a quality assurance plan for periodic testing.

7.11 COMMISSIONING OF A NEW X-RAY BEAM DOSE CALCULATION ALGORITHM

Problem:
Your department has recently installed a major update to the radiotherapy treatment planning system. A part of this update includes the installation of a brand-new X-ray beam dose calculation algorithm. The intention is that this new algorithm will be used for dose calculations for IMRT and VMAT plans. Describe the beam data and parameters that may need to be input into the TPS to commission this new algorithm into clinical service.

Solution:
The commissioning of a new TPS algorithm must follow the methodology published in the IAEA TRS 430 report called 'Commissioning and Quality Assurance of Computerized Planning Systems for Radiation Treatment of Cancer'. This would be the basis for your commissioning.

The beam data that is input into the TPS would be determined by a review of the documentation from the vendor including any reference papers. The actual beam data may consist of any of the following for a variety of geometries such as SSD = 100 cm (or SSD = 90 cm):

- Percentage depth doses

- Cross- and in-plane profiles at a range of depths such as D_{max}, 10, 20 and 30 cm

- Diagonal profiles

- Total scatter factors, phantom scatter factors and/or collimator scatter factors

- Small-field relative output factors

- MLC transmission

- Jaw transmission

- MLC dynamic leaf gaps

- X-ray beam spot size

- Reference dose calibration factor – to relate the monitor units on the linear accelerator to dose in Gy under reference conditions.

Note: Note that as this is being used only for IMRT and VMAT plans, there would not be any fixed or dynamic wedge data measured.

7.12 COMMISSIONING 2D/3D DETECTOR ARRAY

Problem:
Pre-treatment quality assurance is commonly performed using either 2D/3D detector arrays. The detectors either small-volume ionisation chambers or diodes mounted within a large phantom. List the set of tests you would perform to commission one of these devices into clinical use to ensure they are accurate and fit for purpose.

Solution:
The following list of tests are some of those that would be required:
Dosimeter characteristics:

- Physical size of detectors

- Spatial resolution – spacing of the detectors

- Sensitivity of the detectors including minimum and maximum dose ranges

- Total size of the detector which may limit the maximum size of the fields that can be tested

- Limitations of not irradiating any part of the device which is indicated as such with warning notes, e.g., electronics

Calibration process:

- Dose calibration against your reference dosimeter

- Dose calibration factors for all of the radiation beams used, e.g., 6 MV, 6 MV FFF and 10 MV X-ray beams would all have different calibration factors

- Detector array calibrations

Dosimetric performance:

- Warm-up process

- Check background subtraction for system noise

- Beam energy and dose rate variations in response

- Temperature dependence

- Variations in angular response – particularly important for planar detectors

- Short- and long-term stability

- Temperature and pressure corrections in the dosimeter

Clinical testing:

- Correct set-up of software used for analysis, e.g., gamma analysis calculations.

- Measurement comparison for simple and complex fields with ionisation chambers and 2D detectors such as radio-chromic film.

- Ideally one would start with simple open fields such as $10 \times 10 \, cm^2$ and also the largest possible field sizes. Then proceed to test multiple field arrangements with increasing complexity, e.g., open field 4-field box, wedge fields, MLC shaped fields and standard IMRT/VMAT fields.

- Perform tests to check for detection of known errors such as detector offsets in the three axes.

Training and documentation:

- Provide training to all staff in the safe operation of the dosimeter.

- Ensure safe operation in terms of electrical safety and manual handling to minimise injury.

7.13 QUALITY ASSURANCE OF IMRT/VMAT PLANS

Problem:
Describe the measurement techniques by which IMRT/VMAT plans can be checked for accuracy and deliverability.

Solution:
IMRT and VMAT treatment fields made up of a large sum of many smaller fields which when combined provide the required dose distribution. A measurement of the modulated treatment field provides assurance on the accuracy and is one key step in the quality assurance programme. The modulated field is mapped over onto the appropriate phantom and the dose recalculated within the phantom without making any changes to the treatment plan. Accuracy dose measurement provides assurance that the patient treatment plan is correct. Appropriate pass rates for the measurements should be taken from published recommendations and take into account the local technology. The results should be reported for each field but can also be reported for all fields.

Each treatment field should be checked independently and provides:

- Point dose measurements using a small-volume ionisation chamber in a solid water equivalent block phantom at a point corresponding to within the PTV. Suggested tolerance per treatment field is ±3% and for the sum of all fields is ±2%.

- 2D/3D dose arrays – gamma analysis pass rate of 95% at 3%/3 mm or 3%/2 mm.

- Radiochromic film measurements in solid water equivalent block phantoms in one or more dose planes through the PTV – gamma analysis pass rate of 95% at 3%/3 mm or 3%/2 mm.

- Fluence or dose checks using the EPID on the linear accelerator – gamma analysis pass rate of 95% at 3%/3 mm or 3%/2 mm gamma criterion.

Other dosimeters but which are less commonly used include transmission detectors which mounted on the collimator head and 3D gel dosimeters which would be typically only used for commissioning tests.

7.14 CHECK OF THE COLLIMATOR CROSS-HAIR WALKOUT

Problem:
Describe the test to check the walkout of the collimator cross-hair.

Solution:

- Set the gantry angle to 0°, raise the couch-top to isocentre height (SSD of 100 cm) and tape a piece of paper to the couch securely.

- Set the collimator angle to 90° and mark the centre of the cross-hair on the piece of paper.

- Now rotate the collimator by 180° around to a collimator angle of 270° and again mark the centre of the cross-hair on the piece of paper.

- Measure the magnitude and direction of the variation in the two different cross-hair positions.

- Repeat the test but at an extended SSD of, say, 120 cm.

- Tolerance should be 1 mm for linear accelerators including those used for IMRT, SABR and SRS according to the AAPM TG 142 report.

Note: The actual gantry angles used can be varied according to the linear accelerator model being tested.

7.15 CHECK OF THE GANTRY ANGLE READOUT

Problem:
Describe the test to check the accuracy of the gantry angle readout.

Solution:

- This test should be ideally performed with a digital level with an accuracy of better than 0.1°. However, a level device with liquid bubble can also be used.

- The level device will be placed against the gantry reference surface – this is the reference surface as specified by the linac vendor and is typically one of the metal plates on the bottom of the collimator near the cross-hairs.

- Rotate the gantry counterclockwise around to 180° so that the gantry is facing upwards. Check the true gantry angle of 180° against the gantry angle displayed on the linac.

- Rotate the gantry around clockwise to the angles of 270°, 0°, 90° and 180° and repeat the measurement.

- Tolerance is set at 1° according to the AAPM TG 142 report. However, the experience of the authors is such that modern accelerators can easily be accurate to within 0.2°.

Note: The actual gantry angles used can be varied according to the linear accelerator model being tested and which angle scale system being used such as IEC.

7.16 LINEAR ACCELERATOR LASER CHECKS

Problem:
Describe the checks that would be needed to be performed on a monthly basis for the laser system located in the linear accelerator. List the expected tolerances for the lasers.

Solution:
The linear accelerator has lasers which are located on side walls, the ceiling and a sagittal laser which is mounted on the wall facing towards the gantry. All of these lasers must be correctly aligned to the isocentre of the linear accelerator as well as being orthogonal to the axes of gantry and collimator rotation. Traditionally in the past, the isocentre chosen for the lasers was taken to be the mechanical isocentre. More recently with the greater use of IGRT, the radiation isocentre may be chosen as the reference isocentre. Linear accelerators that are designed to be used for SRS have a much tighter mechanical tolerance and the radiation isocentre is determined using the Winston-Lutz test.

One possible set of tests for the lasers is outlined below. It should be noted that there are alternative test methods which are also valid.

Wall lasers:

- Check that the vertical wall lasers are coincident with the gantry axis of rotation and the plane of gantry rotation.

- Check that the wall lasers are truly vertical and horizontal – this can be done using a laser level device, a clear water house or plumb bob device.

- Check that the wall lasers are coincident with each other at the isocentre and over a range at least ±20 cm in each direction from the isocentre position.

Ceiling lasers:

- Check that the ceiling laser is orientated to be truly vertical – this can be done by using a container of water on the floor and checking the coincidence of the reflection with the laser lights.

- Check that the ceiling laser cross-hair centre is coincident with the isocentre.

- Check that the lateral ceiling laser is coincident with the plane of the gantry rotation – note that some linear accelerators will have some noticeable drift/sag as the gantry is rotated around.

- Check that the lateral ceiling laser line is coincident with the side wall lasers.

- Check the coincidence of the longitudinal ceiling laser with the gantry axis of rotation.

Sagittal laser:

- Check that the sagittal laser is truly vertical and not rotated.

- Check that sagittal laser passes through the isocentre.

- Check the coincidence of the sagittal laser with the longitudinal ceiling laser – this should be done above, below and at the isocentre height.

Laser alignment with axes of rotation and isocentre should be within ±1 mm.

The AAPM TG142 report lists three tolerances according to the type of linear accelerator:

Non-IMRT linac: ±2 mm

IMRT linac: ±1 mm

SRS/SBRT linac: <± 1 mm.

7.17 COINCIDENCE OF LIGHT FIELD WITH RADIATION FIELD

Problem:
Describe a method of checking the coincidence of light field with the radiation field including the tolerance of this test.

Solution:
It is noted that there are various methods by which this test can be performed such as using radiochromic film, the EPID panel or a plate containing fluorescent material that glows with light when exposed to ionising radiation.

The following test involves the use of radiochromic film designed for imaging QA.

- Set the gantry angle to be 0.0° and the collimator angle to be 0.0°.

- Place a piece of the radiochromic film onto the couch topic and set the SSD to be 100 cm to the top of the film.

- Set the required field size, e.g., 10×10 cm and using a fine tip pen or marker along with a ruler, mark the middle of the light edge – note that the light edge will have a penumbra and the middle of this should be marked.

- Place a solid water equivalent block on topic of the film – the thickness of the block should be such that is enough to provide maximum electronic build-up, e.g., 1.5 cm for 6 MV X-ray beam or 2.5 cm for a 10 MV X-ray beam.

- Depending on the type of radiochromic film being used, deliver sufficient dose to ensure sufficient darkening of the film and the radiation field edges can be clearly seen.

- Check the coincidence of the radiation field edges with the marks that indicate the light field edge.

Tolerances:

AAPM TG142 tolerances for monthly tests:

2 mm or 1% for symmetric fields.

1 mm or 1% for asymmetric fields – this includes checks with the collimator jaw set to the centre, i.e., jaw position of 0.0 cm.

7.18 TESTING OF THE ISOCENTRE OF THE PLANAR KILOVOLTAGE IMAGING SYSTEM

Problem:
On-board imaging devices make use of a planar X-ray imaging system which delivers low-energy X-ray for patient set-up. Describe the test you would perform to check that the isocentre of the imager is aligned with the isocentre of the linear accelerator. List the tolerances you would use.

Solution:
Test 1

Place a suitable cube phantom on the couch of the linear accelerator, ensure that that the couch is levelled and align to the lasers using the cross-hair lines engraved on the phantom surface. Note that these phantoms typically contain a number of markers which are radiopaque which means they have a different contrast. The simplest version of these phantoms is a small plastic block with a small metal ball located at the centre of the phantom.

Take planar images using the on-board imager at the four gantry angles of 0°, 90°, 180° and 270°. From each of these four images, calculate the difference between the centre of the marker and the digital graticule (also known as the electronic cross-hair). This digital graticule is taken to be correctly aligned to the isocentre of the linear accelerator.

Test 2

An alternative version of the test involves using a CT data set of the same phantom. Align the phantom to the lasers and take a note of the couch coordinates (in cm) in the lateral, longitudinal and vertical axes. Following a similar procedure as in the first test, images are taken for a range of gantry angles. Online image matching is performed using the imager software to match the images of the metal marker in each gantry angle. Once the best agreement is found for all images, a couch shift is performed to the corrected position. The difference in the couch positions is calculated. In addition, the shift in the phantom relative to the laser lines is also noted.

Tolerances:

For conventional linear accelerators, tolerance is ≤2 mm.

For linear accelerators that are used to deliver SABR or SRS treatments, the tolerance should be <1 mm. A tighter tolerance is used here due to the smaller fields used in these treatments.

7.19 SAFETY CHECKS ON LINEAR ACCELERATOR

Problem:
List the safety checks that you would perform on a monthly basis on the multimodality linear accelerator which can generate 6, 10 and 18 MV X-ray beams and electron beams with energy ranging from 4 to 20 MeV.

Solution:
The following items are some of the checks that would be a part of the monthly quality assurance tests. It is noted that some of these should also be tested on a daily basis.

- Check for any mechanical issues with the linear accelerator such as scratches or cracks.

- The functionality of the audio-visual system ensuring the camera and audio systems are working.

- That all of the beam-on signs work correctly when any radiation beam is switched on.

- That the X-ray beam-on sign lights up when the on-board imager device is being operated.

- Check the last-person out system works correctly including the audible alarms.

- Check the functionality of the sensors with the entrance to the maze is working correctly – these are typically IR curtains or motion sensors.

- Check the functionality of any physical barriers into the maze such as a gate.

- Check that the neutron door is working correctly and that the 18 MV X-ray beam cannot be switched on if the door is open.

- Verify that the beam switches off if an object is pushed into the maze, e.g., can be done with a rolling chair.

- Ensure that touch-guards on the linac including the on-board imaging system are functioning correctly.

- Check that all of the beam modifiers are in good working order and no apparent physical damage or defects, e.g., fixed wedges, electron applicators and patient immobilisation devices.

- Review the linear accelerator logbook to check for any major issues or repairs performed by the service engineers.

- A check of the emergency off buttons located on the linear accelerator, treatment couch, inside the bunker and at the console region all function correctly. However, it is important to note that this particular test can cause possible issues in terms of having a major power cut to the linear accelerator. As such, this test should be done in consultation with the service engineers or at a reduced frequency.

7.20 RECORD AND VERIFY SYSTEM

Problem:
Describe the purpose of a radiotherapy record and verification system.

Solution:
A radiotherapy record and verify (R&V) system is a computerised database which connects the linear accelerators, treatment planning systems and imaging devices such as CT and MRI scanners.

This database contains the DICOM information for each radiotherapy patient such as the treatment plan parameters, 3D dose calculations and imaging scans. The R&V system also has the parameters for each of the linear accelerators.

With increasing complexity of radiotherapy plans such as IMRT and VMAT, their delivery for the patient cannot be done manually. For this reason, R&V systems are required to control the linear accelerator during the treatment with all settings such as beam modality, energy and dose rate; monitor units; gantry, collimator and couch angles; couch positions; collimator jaw positions and MLC positions. These settings for the patient plan are determined during the planning process.

The treatment staff will load up the plan information for the individual patient and this will be used during the set-up of the patient on the treatment couch. If any of the settings on the linear accelerator do not agree, within accepted tolerances, the issue needs to be resolved before treatment can commence.

The R&V system stores all of the pre-treatment imaging taken for patient alignment which can be retrospectively reviewed to ensure accurate set-up as well as to monitor for changes in the patient shape. The R&V system also records the delivered dose to the patient which is an important for the patient records.

CHAPTER 8

Brachytherapy Treatment and Quality Assurance

8.1 RADIONUCLIDES FOR BRACHYTHERAPY

Problem:
List the important features that gamma-emitting radionuclides should have so that it would be useful for brachytherapy applications.

Solution:
Some of the key physical design features of the radionuclides are as follows:

- The radionuclide should have a high specific activity so that it emits enough radiation to provide a therapeutic dose to the patient. It also means that the radionuclide can be made from a smaller size source.

DOI: 10.1201/9780429159466-8

- The emitted energy should be such that is provides sufficient penetration in the tissue but not so high as to too penetrative and too much dose is deposited outside of the treatment area.

- The half-life should be suitable for the particular purpose – a high dose rate (HDR) treatment unit requires a radionuclide with sufficiently long life so that the source does not need to be replaced too often. For seed brachytherapy, a shorter half-life is required to minimise radiation safety issues for the patient.

- The radionuclide should be able to be fabricated so that the source is appropriately encapsulated inside the metal casing (typically steel or titanium).

- No toxic gases should be produced in the radionuclide decay.

- Any beta rays emitted by the radionuclide during the decay process should ideally be absorbed in the metal encapsulation.

- The radionuclide should not be toxic in case there is leakage of the source either into the patient or if handled by the physicist when being prepared for use.

8.2 CHARACTERISTICS OF LOW-DOSE RATE AND HIGH-DOSE RATE BRACHYTHERAPY

Problem:
Brachytherapy can be classified according to their dose rates. Compare the characteristics of low-dose rate (LDR) and HDR brachytherapy in terms of dose rate, activity and treatment time. Compare the advantages and disadvantages of both techniques, assuming they are performed using remote afterloading equipment.

Solution:

Characteristics	LDR	HDR
Dose rate	0.2–2.0 Gy/h	>12 Gy/h
Activity	3.7 GBq (100 mCi)	370 GBq (10 Ci)
Treatment time	Long treatment time (hours/ days/permanent implants)	Short treatment time (minutes) and delivered in multiple fractions

8.3 SEALED VERSUS UNSEALED SOURCES

Problem:

Write short notes comparing sealed and unsealed source. Give two examples for each category of the sources.

Solution:

Sealed source:

- Sealed source or sealed radionuclide is a radiation source consisting of any radioactive material, nuclear material or prescribed substance firmly incorporated in solid and inactive material or sealed in an inactive container of sufficient strength (usually stainless-steel) to prevent, under normal conditions of use, any dispersion of its contents.

- During brachytherapy, sealed sources are implanted/positioned within or at close proximity to the target volume.

- Sealed sources when intact pose external radiation hazards.

- Examples of sealed source are Co-60 and Ir-192.

Unsealed source:

- Unsealed sources consist of powders, liquids or sometimes gases that contain radioactive elements and that could easily be released from their containers through leaks and spillages and dispersed into the environment.

- When given orally or intravenously unsealed sources localise to the target tissue by virtue of its biological, physical or chemical property.

- Unsealed sources pose internal and external radiation hazards.

- Examples of unsealed sources are I-131 and P-32.

8.4 HDR REMOTE AFTERLOADING

Problem:
Provide several reasons why HDR remote afterloading systems are replacing manual afterloading iridium wire for interstitial implants? Are there any disadvantages?

Solution:
Advantages of HDR remote afterloading systems:

- Elimination or reduction of exposure to medical personnel.

- Capability of optimising dose distribution beyond what is possible with manual afterloading.

- Treatment technique can be more consistent and reproducible.

- Permits treatment on an outpatient basis, using multiple fractions regiments.

Disadvantages could be a lack of normal tissue sparing in regard to late effects due to the high dose per fraction. In addition, more quality assurance is required, and the high activity of the source may lead to security concerns.

8.5 SOURCE ENCAPSULATION

Problem:
Give the reasons why brachytherapy sources are encapsulated within a metal casing.

Solution:

Brachytherapy sources are usually sealed so that the radioactive material is contained fully encapsulated within a protective capsule. With encapsulation, radioactive sources are less likely to be damaged under any foreseeable degree of physical or chemical stress. This capsule is also designed to prevent leakage or escape of the radioactive source and it makes the source rigid. Contamination of applicators and transfer tubes is avoided. In addition, encapsulation provides filtration of unwanted alpha or beta radiation.

8.6 IR-192 AS BRACHYTHERAPY SOURCE

Problem:

Ir-192 is a common radioactive source used for brachytherapy.

a. State the decay mechanism of Ir-192, its half-life and average energy of gamma-ray released during the decay.

b. Give two reasons why Ir-192 is a suitable radioactive source for brachytherapy.

c. Suggest one practical reason why Ir-192 is a less than ideal radioactive source in HDR afterloading equipment.

Solution:

a. 95% of the time Ir-192 decays through negative beta emission to Pt-192. The remaining 5% of the time Ir-192 decays through electron capture to Os-192. It has a half-life of 74 days. It emits gamma-ray with an average energy of 0.35 MeV.

b. For brachytherapy with sealed sources, Ir-192 is a pure gamma emitter. It also has high specific activity, or activity per unit mass, which means that a very small source can result in high dose rate and hence shorter treatment times.

The effective photon energy of around 0.35 MeV ensures a sufficient absorbed dose at a sufficient distance from the source to treat the target homogeneously.

c. Ir-192 has a relatively short half-life (74 days) and requires a change of source every 3–4 months which is costly, making it less than ideal radioactive source in HDR afterloading equipment.

8.7 SOURCE DECAY (1)

Problem:
An Ir-192 source has an initial activity of 13.5 Ci. Calculate the activity of the source after 2 months. (Given $T_{1/2}$ for Ir-192 = 74.02 days and decay constant, $\lambda = 0.693/T_{1/2}$.)

Solution:
Decay constant, $\lambda = \dfrac{0.693}{T_{1/2}}$

$$= \frac{0.693}{74.02}$$

$$= 9.36 \times 10^{-3} \text{ days}^{-1}$$

Use the decay equation $A = A_o e^{-\lambda t}$ where

A = ending activity

A_0 = initial activity

t = elapsed time

We get the activity after 2 months (60 days),

$$A = (13.5) e^{\left(-9.36 \times 10^{-3}\right)(60)}$$

$$= 7.7 \; Ci$$

8.8 SOURCE DECAY (2)

Problem:

A Geiger counter detects 40 counts per second from a sample of Iodine-131. The half-life of I-131 is 8 days. Using the axes given in Figure 8.1, sketch a curve showing the count rate from the sample of I-131 over a period of 20 days. Show your calculations.

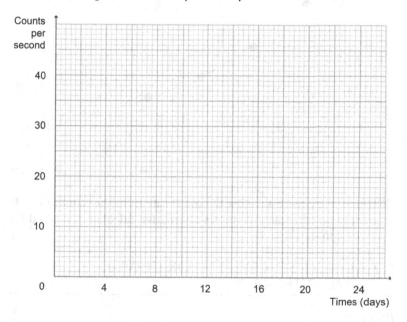

FIGURE 8.1 A graph sheet for sketching count rate from a sample of I-131 over a period of 20 days.

Solution:

Decay constant, $\lambda = \ln 2/8 = 0.087\,\text{day}^{-1}$.

Counts per second after t days, $A_t = A_0 e^{-\lambda t}$.

After 4 days, $A_4 = 40e^{-(-0.087 \times 4)} = 28$ counts per second.

Repeat the above for $t = 8$, 12, 16 and 20.

We obtain $A_8 = 20$ counts/s, $A_{12} = 14$ counts/s, $A_{16} = 10$ counts/s and $A_{20} = 7$ counts/s.

The results can now be plotted on the graph as below as shown in Figure 8.2.

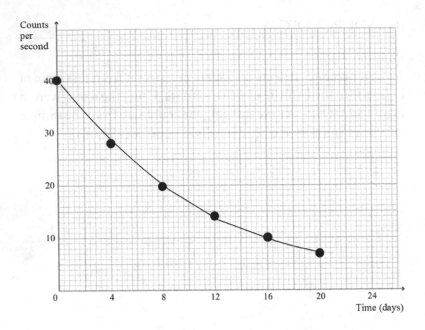

FIGURE 8.2   A curve showing the count rate from the sample of I-131 over a period of 20 days. *Note:* An alternative calculation method for $A_8 = 40/2 = 20$ counts/s.

8.9 MANCHESTER SYSTEM FOR CERVIX BRACHYTHERAPY

Problem:

a. The Manchester system for cervix brachytherapy is characterised by doses to points A and B. Define these points.

b. Draw a labelled diagram showing a typical dose distribution following the Manchester system for cervix brachytherapy. Show the location of points A and B, uterine tandem and pair of ovoid in your diagram.

c. At which point is the dose prescribed and why?

d. Explain the disadvantage of using the point indicated in your answer in (c) for prescription of dose for brachytherapy.

Solution:

a. Point A is defined as a point 2 cm lateral to the centre of the uterine canal and 2 cm superior to the mucosa of the lateral fornix, in the plane of the uterus. Point B is defined 2 cm above external os and 5 cm laterally to midline.

b. See Figure 8.3.

c. The dose is prescribed to point A. Point A represents the location where the uterine vessels cross the ureter. It is believed that the tolerance of these structures is the main limiting factor in the irradiation of the uterine cervix.

d. The exact meaning and their definitions have not always been interpreted in the same way in different centres and

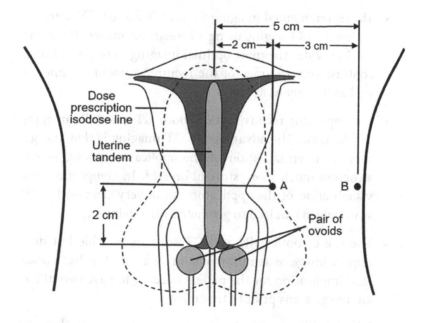

FIGURE 8.3 A diagram showing a typical dose distribution following the Manchester system for cervix brachytherapy with the location of the points A and B, uterine tandem and pair of ovoids.

even in the same centre over a period of time. The different methods of definition provide different values for the calculated dose rate to point A.

8.10 IMAGING FOR CERVIX BRACHYTHERAPY

Problem:
High dose rate brachytherapy can be used for the treatment of cervical cancer. The brachytherapy is often delivered to the patient in addition to external beam radiotherapy. Discuss the different imaging modalities that may be used for (i) determining the GTV and CTV volumes and for (ii) reconstruction of the treatment applicators.

Solution:

- The recommended imaging for the GTV and CTV determination is MR imaging using T2 weighted images. The use of MRI provides the most optimal imaging in terms of image contrast and accuracy of the required tumour volumes as well as the organs at risk.

- For applicator reconstruction, both CT and MRI imaging can be used. The advantage of CT imaging is that you get very good visualisation of the applicator making reconstruction much more straightforward. In comparison, the visualisation of the applicators is not very clear with MRI scans which can lead to geometric uncertainties.

- The use of both MRI and CT scans is possible but does require image registration of both data sets. This has a practical implication for the patient needing to have two different image scans prior to treatment.

- Ultrasound imaging is mainly used to ensure that the applicators are inserted correctly into the patient and in verification of the applicator position particularly if the patient has to be moved around for imaging. However, ultrasound can also be used for treatment planning and verification of the target volumes.

8.11 INTERSTITIAL BRACHYTHERAPY

Problem:
The Paris system for interstitial brachytherapy is used primarily for single- and double-plane implants.

a. List five general rules for placement of the sources to achieve the desired dose distributions.

b. How is the reference dose rate determined in the Paris system?

Solution:

a. Five general rules for placement of the sources based on the Paris system:

- Sources must be linear and their placement parallel.

- Centres of all sources must be located in the same plane (central plane).

- Linear source strength (activity) must be uniform and identical for all sources.

- Adjacent sources must be equidistant from each other.

- Spacing between sources should be wider when using long sources.

b. The reference dose rate is a fixed percentage (85%) of the basal dose rate. The basal dose rate is the average of the minimum dose rates located between the sources inside the implanted volume.

8.12 WELL CHAMBER CALIBRATION OF IR-192 SOURCES (1)

Problem:
Describe how a well chamber is used for the calibration of Ir-192 brachytherapy sources.

Solution:
A well-type ionisation chamber is used to measure the source strength measurements for all kinds of brachytherapy sources. This is a large volume ionisation chamber typically 250 cm³ into which the source is positioned inside at the mid-point by connection with transfer tubes from the HDR unit to the well chamber. The positioning is important as the reading from the chamber will be a maximum at the mid-point and this position should be used to better than ±5 mm. The measurement is performed by either measuring the current at this point or measuring the charge per unit time, e.g., nC per second.

The air-kerma strength of the source S_k is given by the following equation:

$$S_k = I \times k_{\text{TP}} \times k_{\text{elec}} \times N_k \times A_{\text{ion}} \times P_{\text{ion}}$$

here

I = current reading (A)

k_{TP} is the temperature and pressure correction factor and is given by

$$k_{TP} = \left(293.2 + \frac{T(^\circ C)}{293.2}\right) \times \left(\frac{P(m\text{Bar})}{1013.25}\right)$$

k_{elec} is the electrometer calibration factor and usually has a value of 1.000,

N_k is the chamber calibration factor in terms of air-kerma strength per unit current cGy m²h⁻¹,

A_{ion} is the correction for collection efficiency at the time of calibration,

P_{ion} is the ionic recombination correction factor and typically has a value close to 1.000.

8.13 WELL CHAMBER CALIBRATION OF IR-192 SOURCES (2)

Problem:
Calculate the air-kerma strength of an Ir-192 brachytherapy source using a well ionisation chamber? Assume that the source is correctly located at the sweet spot of the chamber and the following measurement values are provided.
Air temperature=23.3°C
Air pressure=1014.5 mBar
Charge readings collected for 1 minute each: 1539.9 nc, 1539.9 nC, 1540.2 nC, 1540.2 nC and 1539.9 nC.
The well chamber has a calibration certificate with N_k value of 9.267×10^7 cGy m^{2}h^{-1} A^{-1}.
The electrometer calibration certificate provides an electrometer calibration factor k_{elec} value of 1.000.
The correction for collection efficiency at the time of calibration is 1.000.

Solution:
Using the equation for the air-kerma strength of a brachytherapy source, S_k, is given by the following equation:

$$S_k = I \times k_{TP} \times k_{elec} \times N_k \times A_{ion} \times P_{ion}$$

I=average reading/second=2.567×10^{-8} A

$$k_{TP} = \left(293.2 + \frac{T(°C)}{293.2} \right) \times \left(\frac{P(mBar)}{1013.25} \right) = 1.010$$

P_{ion} and A_{ion} have a value of 1.000
Using the equation, gives a source strength=2.401 cGy m^{2}h^{-1}.

8.14 AAPM TG43 DOSE FORMALISM

Problem:
Describe the AAPM TG 43 dose calculation formalism for brachytherapy sources which is used to calculate a 2D dose distribution. Explain the meaning of the different functions used in the formalism.

Solution:
The dose rate $\dot{D}$ (r, θ) from a brachytherapy source calculated at a distance r and an angle θ is given by the following equation:

$$\dot{D}(r,\theta) = S_k \Lambda \, \frac{G(r,\theta)}{G(r=1 \text{ cm and } \theta = 90°)} \, g(r) \, F(r,\theta)$$

where

S_k is the air-kerma strength of the source with units of cGy cm^2 hr^{-1}.

Λ is the dose rate constant in water for the source defined at a distance of 1 cm from the source and on the bisector at 90° from the source. Λ is defined in units of cGy hr^{-1} per unit air-kerma strength. The value of this constant will be provided by the vendor and varies according to the particular isotope and design of the brachytherapy source.

$G(r, \theta)$ is the geometry function and describes how the dose varies as a function of distance from the source. This function is normalised to the dose at a point that is 1 cm distance and at an angle of 90° from the source. This function has units of cm^{-2}.

The function $g(r)$ is defined as the radial dose function and takes into account the absorption and scattering of the radiation within the medium of interest along the transverse axis of the source. The function is normalised to the dose at a distance of 1 cm from the source.

The function $F(r, \theta)$ is the anisotropy function and takes into account the absorption and scattering of the radiation within the medium as well as in the source encapsulation.

8.15 HDR BRACHYTHERAPY QA CHECKS

Problem:
List the daily QA checks you would perform on a HDR brachytherapy unit which contains either an Ir-192 or Co-60 source.

Solution:
The daily QA checks would include the following:

- Independent radiation monitor is working correctly by detecting radiation when the source is out.

- Radiation survey meter functioning correctly.

- All safety interlocks are working as expected:

  - Treatment interruption and emergency retraction buttons immediately retract the source.

  - The source cannot be sent out if the door interlock is not set, key switch not turned (if present), missing transfer tubes, applicators are not connected and the applicator locking ring is not locked into place.

- The treatment unit displays the correct date and time.

- The current source strength is compared against an independent calculation such as on written table or external spreadsheet.

- Testing of the source position accuracy and dwell time accuracy – this can be done using a source position ruler check device.

- The source retracts into the treatment safe at the end of the previous test.

- All applicators, transfer tubes and the HDR unit do not have any visible damage.

- Audio visual system is working correctly.

- The emergency equipment is readily accessible in case of stuck source: large lead pot, large tweezers and wire cutters.

CHAPTER 9

Basic Radiobiology

9.1 RADIATION EFFECTS (1)

Problem:
The effects of radiation can be categorised into stochastic effects and tissue reactions. Explain what is meant by the terms stochastic effects and tissue reactions, and give two examples of each type of effect.

Solution:
Stochastic effects are probabilistic effects that occur by chance. The probability of occurrence is typically proportional to the dose received. Stochastic effects after exposure to radiation occur many years later (the latent period). The severity is independent of the dose originally received. Examples of stochastic effects are cancer and inherited defect in offspring.

Tissue reactions (previously known as deterministic effects) are short-term, adverse tissue reactions resulting from a dose that is significantly high enough to damage living tissues. The severity of a tissue reaction increases with radiation dose above a threshold, below which the detectable tissue reactions are not observed. Tissue reactions are usually predictable and reproducible. For example, localised doses to certain parts of the body at increasing

DOI: 10.1201/9780429159466-9

levels will result in well-understood biological effects. Examples of tissue reactions are cataract and skin burn to erythema.

9.2 RADIATION EFFECTS (2)

Problem:
There has been a lot of discussion about the linear no-threshold model for determination of radiation risks at low doses. Why is the scientific information on radiation effects of low radiation doses (e.g., dose < 20 mSv) limited?

Solution:

- Dosimetry is difficult at low level of radiation – close to background radiation.

- Limited epidemiological evidence due to very large number of people required to demonstrate the effect.

- Research and experiments with humans are ethically impossible.

- The radiation effects are too small (if any).

- It is likely that there is a dose and dose rate effect – at lower doses and dose rates radiation effects are likely to be smaller than at high doses.

9.3 EARLY VERSUS LATE RESPONDING TISSUE

Problem:
Compare 'early' versus 'late' responding tissues during a course of radiotherapy treatment by completing the following table:

Characteristics	Early Responding Tissues	Late Responding Tissues
Proliferation rate		
Onset for occurrence		
Type of tissue damage		
Sensitivity to fractionation		
Effect of overall treatment time		
Sites/structures of damage		

Solution:

Characteristics	Early Responding Tissues	Late Responding Tissue
Proliferation rate	Rapidly proliferating	Slowly proliferating
Onset for occurrence	Onset < 90 days	Onset > 90 days, typically 0.5–5 years
Type of tissue damage	Transient, but consequential late reactions may occur	Irreversible
Sensitivity to fractionation	Low sensitivity to fractionation	High sensitivity to fractionation
Effect of overall treatment time	Shorter overall treatment time causes greater injury	Less effect of overall treatment time
Sites/structures of damage	Rapidly proliferating	Slowly proliferating

9.4 SERIAL VERSUS PARALLEL ORGANS

Problem:

a. Define serial and parallel organs. Give an example for each organisation.

b. How does the difference in normal tissue organisation affect tolerance doses for small and large volume irradiation?

Solution:

a. Anatomical structures are composed of functional subunits (FSUs) and an FSU is the largest tissue volume or unit of cells that can be regenerated from a single surviving clonogenic cell. Serial organs have their FSUs arranged in series. The function of the entire organ depends on the function of each individual FSU, e.g., spinal cord. Parallel organs have their FSUs arranged in parallel to other FSUs such that they function independently of each other, e.g., lungs.

b. Parallel organisation consists of a large number of FSU. Many FSUs can be inactivated without leading to loss of organ function. Functional damage will not occur until a critical number of FSUs are inactivated by radiation. There is a threshold volume of irradiation below which no functional damage will develop, even after high dose of radiation. Above this volume threshold, damage occurs as a graded response of increasing severity with increasing dose. This large reserve capacity increases the tolerance to partial volume irradiation.

For serial organisation, the integrity of each FSU is critical to organ function. The inactivation of one subunit may cause loss of function in the whole organ. Radiation damage is expected to show binary response with a threshold dose below which there is a normal organ function and above which there is loss of function. The probability of inactivation of a particular subunit by a given radiation dose increases with increasing length of tissue irradiated, but there is no threshold volume for the development of the complication endpoint.

9.5 HEALTH RISK MODELS FROM RADIATION EXPOSURE

Problem:

a. Describe briefly the following models of health risk from radiation exposure:

   i. Linear no-threshold (LNT) model

   ii. Threshold model

   iii. Hormesis model

b. Draw three curves of health risk versus dose on the graph below representing the models as indicated in questions (a) (Figure 9.1).

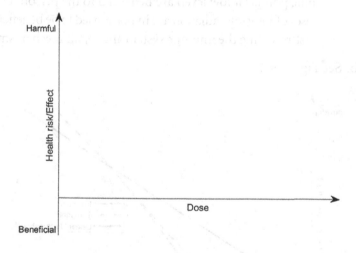

FIGURE 9.1 A graph of health risk/effect versus dose.

Solution:

a.

    i. The LNT model is a model used in radiation protection to estimate the long-term, biological effect caused by ionising radiation. It assumed that the risk or damage due to radiation is directly proportional ("linear") to the dose of radiation, at all dose levels. The LNT risk model is the current human health risk assessment paradigm.

    ii. Threshold model assumes that radiation has no health effect up to a certain dose. After that dose, health effect from radiation exposure may be observed and the severity/risk of the effect increases with dose.

    iii. Hormesis model assumes that radiation in high doses increases the incidence of cancer; however, doses up to some points (at low level) are beneficial to the person. Low doses of ionising radiation are hypothesised to be beneficial by stimulating the immune system and repair mechanisms.

b. See Figure 9.2.

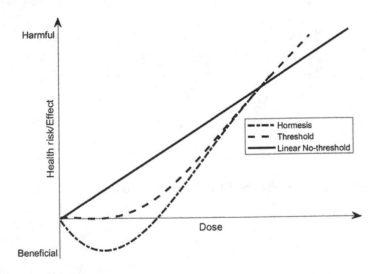

FIGURE 9.2 The curves of radiation health risk versus dose for (i) linear no-threshold model, (i) threshold model and (iii) hormesis model.

9.6 DIRECT ACTION VERSUS INDIRECT ACTION OF RADIATION

Problem:
Explain the differences between cell damage by direct action and indirect action.

Solution:
Figure 9.3 shows a diagram illustrating direct versus indirect action of radiation. In direct action, the radiation interacts directly with the target cell (DNA) and produces biological damage directly. This process is predominant with high LET radiations such as α-particles and neutrons, and high radiation doses. In direct action, the radiation hits the DNA molecule directly, disrupting the molecular structure. Such structural change leads to cell damage or even cell death. Damaged cells that survive may later induce carcinogenesis or other abnormalities. The process is known as direct because the interaction occurs directly between a particle and a cellular component without an intermediary step.

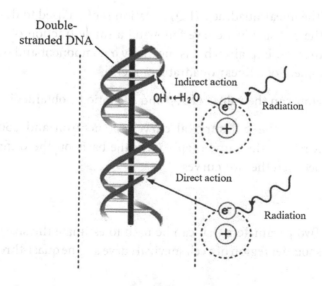

FIGURE 9.3   Direct versus indirect action of radiation.

In the indirect action of radiation, the radiation hits the water molecules, the major constituent of the cell, and other organic molecules in the cell, whereby free radicals such as hydroxyl (HO•) and alkoxy (RO$_2$•) are produced. These free radicals are highly reactive molecules because they have an unpaired valence electron. The free radicals can break the chemical bonds of the DNA and produce chemical changes that eventually produce biological damage. Indirect action damage is mainly caused by low LET radiations and it can be modified by chemical sensitisers or radiation protectors.

9.7 CELL SURVIVAL CURVE

Problem:
A cell survival curve describes the relationship between the absorbed doses of radiation.

a. Based on the 'multi-target theory', what two parameters can be used as estimates of the size of the shoulder region of a cell survival curve? Illustrate the parameters on a cell survival curve.

b. The linear quadratic (LQ) equation is often used to describe the cell survival curve following a single fraction of radiotherapy. Explain what is meant by α component and β component of a linear quadratic model?

c. How can the values of α, β and α/β ratio be obtained?

d. Sketch the cell survival curves for neutron and 250 keV X-ray on the same graph. Give the basis on the difference between the two curves.

Solution:

a. Two parameters that can be used to estimate the size of the shoulder region of a cell survival curve are the quasi-threshold dose (D_q) and the extrapolation number (n). D_q represents the dose in between the 100% survival point and the extrapolation line. n is the intersection between the y-axis and the extrapolated exponential part of the curve (Figure 9.4).

b. α is a linear coefficient which is directly proportional to dose that corresponds to cells that cannot repair themselves (irreversible lethal damage) after a single-hit event. This component is important in high LET radiation. β is a quadratic coefficient which is directly proportional to the square of dose that corresponds to cells that stop dividing after more than one radiation hit. This results in sub-lethal (repairable) damage which can be lethal in combination. The β component is important in low LET radiation.

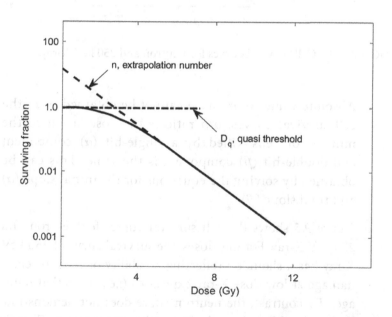

FIGURE 9.4 The quasi-threshold dose (D_q) and extrapolation number (n) can be used to estimate the size of the shoulder region of a cell survival curve.

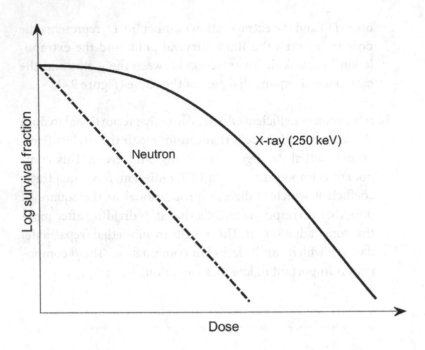

FIGURE 9.5   Cell survival curves for neutron and 250 keV X-ray.

c. Absolute values of α and β cannot be obtained from the cell survival curves. α/β ratio is the dose at which the number of cells killed by a single-hit (α) component and double-hit (β) component is the same. This can be obtained by solving the equations for the initial slope (α) and final slope (β).

d. Figure 9.5 shows the cell survival curves for neutron and 250 keV X-ray. For low doses, the survival curve of 250 keV X-ray has a shoulder: reflecting an ability of cells to repair damage at low dose X-ray exposures (i.e., sub-lethal damage). By contrast, the neutron curve does not demonstrate such a shoulder where the survival curve is linear. There is no shoulder due to the increase of killing by single event by neutron.

9.8 CELL DAMAGE

Problem:

a. Explain briefly three broad types of cellular radiation damage.

b. Figure 9.6 shows the survival of Chinese hamster cells following two fractions of radiation given at different time intervals. These experiments were carried out seven times for seven different time intervals.

    i. Describe the relationship shown in Figure 9.6.

    ii. Name and describe the phenomenon shown by this experiment.

   iii. In a follow-up experiment, increased survival was seen when radiation fractions were more than 12 hours apart. Explain this phenomenon.

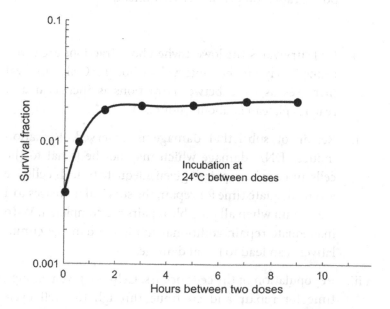

FIGURE 9.6 Survival of Chinese hamster cells following two fractions of radiation given at different time intervals.

Solution:

a. The three types of cellular radiation damage are lethal damage, sub-lethal damage and potentially lethal damage.

- Lethal damage is a type of cellular radiation damage in which the cellular DNA is irreversibly damaged or irreparable to such an extent that the cell dies or loses its proliferative capacity.

- Sub-lethal damage is a type of cellular radiation damage in which partially damaged DNA is left with sufficient capacity to restore itself over a period of a few hours, provided there is no further damage during the repair period.

- Potentially lethal damage is a type of cellular radiation damage in which repair of what would normally be a lethal event is made possible by manipulation of the post-irradiation cellular environment.

b.

i. Cell survival is the lowest when both fractions are given immediately (time interval = 0 hour). Cell survival increases as time between fractions is increased and reaches plateau at about 2 hours.

ii. Repair of sub-lethal damage is observed. Radiation induces DNA damage which may not be lethal to the cells and can be repaired given adequate time. If cells are given adequate time for repair, the survival increases to a maximum when all possible repairs are completed. With inadequate repair, additional sub-lethal damage cumulatively can lead to lethal damage.

iii. Repopulation of the cells occurs. Cells are given enough time for repair and continue through the cell cycle to repopulate due to long interval between radiation fractions.

9.9 α/β RATIO

Problem:

a. What does an α/β ratio value indicate in radiobiology?

b. Compare and contrast the following tissues as stated in the table:

Characteristics	Tissue with Large α/β (>10 Gy)	Tissue with Small α/β (<5 Gy)
Proliferation rate		
Dependence on overall treatment time		
Shape of dose-response curve		
Impact of reducing dose per fraction		
Example of tissue		

Solution:

a. The α/β ratio relates the relative importance of single- and double-strand breaks in causing cell death. In effect, alpha/beta ratio indicates how resistant a cell is to radiation damage. It provides the dose (in Gy) where cell killing from linear and quadratic components of the linear quadratic equation are equal. At points below this value, linear cell kill is dominant; in those after quadratic cell kill takes over. This ratio is used in many areas of radiotherapy.

b.

Characteristics	Tissue with Large α/β (>10 Gy)	Tissue with Small α/β (<5 Gy)
Proliferation rate	In rapidly proliferating tissues, turnover times days or weeks	In slowly proliferating tissues, long turnover times

(Continued)

Characteristics	Tissue with Large α/β (>10 Gy)	Tissue with Small α/β (<5 Gy)
Dependence on overall treatment time	Spared by prolongation of treatment time Increased by shortening of treatment time	Insensitive to overall treatment time
Shape of dose-response curve	Straighter dose-response curve	Bendy dose-response curve
Impact of reducing dose per fraction	Not much sparing by reducing dose per fraction	Spared by smaller dose per fraction
Example of tissue	Most tumours, e.g., head and neck, cervix	Some slow-growing tumours, e.g., melanoma, liposarcoma and prostate

9.10 RE-OXYGENATION

Problem:

a. Define re-oxygenation.

b. With the aid of a graph, describe the process of re-oxygenation in relation to the hypoxic fraction in a tumour following irradiation.

c. List possible mechanisms of re-oxygenation.

d. What is the clinical implication of re-oxygenation?

Solution:

a. Re-oxygenation is a process by which surviving hypoxic clonogenic cells become better oxygenated during the period after irradiation of a tumour, leading to an increase in radiosensitivity.

b. See Figure 9.7.

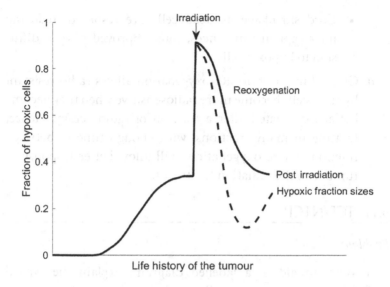

FIGURE 9.7   A curve showing the process of re-oxygenation in relation to the hypoxic fraction in a tumour following irradiation.

Small lesions (1 mm) are well-oxygenated, but as the tumour grows bigger the hypoxic fraction rises. A single dose of radiation kills the radiosensitive oxic cells while the radio-resistant hypoxic cells survive; hence the hypoxic fraction rises steeply to nearly 100%. Re-oxygenation then occurs in the hypoxic cells; they become better oxygenated and the hypoxic fraction falls. Rate of re-oxygenation is variable (a few hours to a few days) in different tumours and microenvironment (e.g., oxygen saturation).

  c. Mechanisms of re-oxygenation include:

   • Recirculation through temporarily closed vessels.

   • Reduced respiration rate in damaged cells.

   • Ischaemic death in cells without replacement.

   • Mitotic death of irradiated cells.

- Cord shrinkage as dead cells are resorbed, reducing inter-capillary distances, thus improved oxygen diffusion to hypoxic cells.

d. Clinical implication: Re-oxygenation allows radio-resistant hypoxic cells become more radiosensitive when they become better oxygenated after a fraction of radiotherapy. Hence, treating in many fractions, with enough time in between fractions for re-oxygenation, will allow better response of tumour to the radiation treatment.

9.11 TCP/NTCP

Problem:

a. With an aid of a labelled diagram, explain the typical relationships between radiation dose with the tumour control probability (TCP) and normal tissue complication probability (NTCP) expected from a radical course of radiotherapy.

b. What is 'therapeutic ratio'?

c. Explain the aim of radioprotector in radiotherapy.

Solution:

a. In radiotherapy, the TCP is a parameter used to calculate the probability of tumour control. A TCP curve is a plot of the probability tumour control (or tumour cells killing) versus dose. This curve is typically sigmoid in shape, with minimal probability of tumour control at low doses, followed by a rapid rise in the tumour control probability once a particular dose is achieved, which slows as dose increases further. An NTCP is a parameter describing the probability of normal tissue complication or damage. The curve is a plot of the probability of normal tissue complication versus dose. It generally follows the shape of a TCP curve but occur at higher level of doses (Figure 9.8).

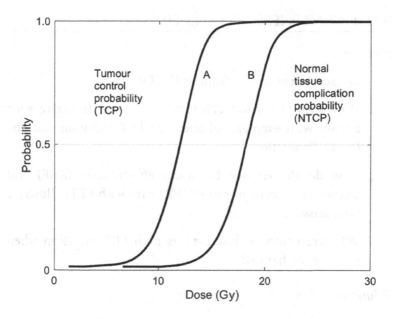

FIGURE 9.8 A typical relationship between radiation dose with TCP and NTCP from a radical course of radiotherapy.

b. Therapeutic ratio generally refers to the ratio of the TCP and NTCP at a specified level of response (usually 0.05) for normal tissue. This concept is often used to represent the optimal radiotherapy treatment.

c. Radioprotectors are compounds used to prevent/protect the non-tumour cells from the harmful effects of radiation. With reference to the TCP/NTPC curves, the aim of using a radioprotector is to move the normal tissue complication curve to a higher dose region by protecting normal cells while not affecting the tumour control curve, or at least not altering it as much. The outcome would decrease the normal tissue complication probability that would be achieved for a given level of tumour control.

9.12 LINEAR ENERGY TRANSFER

Problem:

a. What is linear energy transfer (LET)?

b. How does LET affect cell survival curves? Illustrate your answer with examples of one high LET radiation and one low LET radiation.

c. How do the relative biological effectiveness (RBE) and oxygen enhancement ratio (OER) vary with LET? Illustrate your answer.

d. Why are neutrons described as high LET radiation when they are uncharged?

Solution:

a. LET is the average energy of deposited per unit path length as a charged particle travels through matter by a particular type of radiation. Its units are keV/μm. The LET of a charged particle is proportional to the square of the charge of the particle and is inversely proportional to the kinetic energy of the particle (Figure 9.9) $\left(\text{LET} \propto \dfrac{\text{charge}^2}{\text{Kinetic energy}} \right)$.

b. Low LET radiation results in a cell survival curve with an initial slope followed by a shoulder region and then become nearly straight at higher doses. High LET radiation on the other hand produces a curve that is almost an exponential function of dose, shown by an almost straight line on the log-linear plot Figure 9.9.

c. RBE is defined as the ratio of the doses required by radiation of two different energies to cause the same level of effect. RBE increases with increasing LET reaching a peak at 100 keV/μm then decreases. At 100 keV/μm density of ionisation, the average separation between ionising events just about

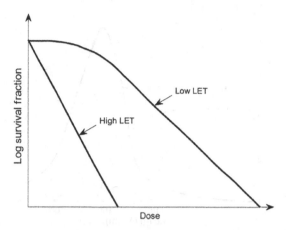

FIGURE 9.9   Cell survival curves with high and low LET radiations.

coincides with the diameter of the DNA double helix (20 Å, or 2 nm). Radiation with this density of ionisation has the highest probability of causing a double-strand break by the passage of a single-charged particle.

Much more densely ionising radiations (with an LET > 100 keV/μm) readily produce double-strand breaks, but energy is 'wasted' because the ionising events are too close together. Because RBE is the ratio of doses producing equal biological effect, this more densely ionising radiation has a lower RBE than the optimal LET radiation. The more densely ionising radiation is just as effective per track, but less effective per unit dose. This is called overkill effect.

OER is the ratio of dose needed without oxygen, to the dose needed with oxygen to have the same effect. The oxygen effect is more pronounced with low LET radiations, resulting in higher OER values which typically range in between 2.5 and 3. As the LET increases, the OER falls slowly at first, after which it falls rapidly and reaches unity. This indicates that oxygen effect is less effective or non-existent with high LET radiations. With an increase in LET, or more densely radiation, OER decreases to unity.

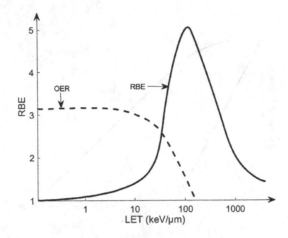

FIGURE 9.10   Cell survival curves with high and low LET radiations.

The two curves virtually mirror each other. The optimal RBE and the rapid fall of OER occur at about the same LET, 100 keV/μm (Figure 9.10).

d. Although neutrons are uncharged and do not directly produce ionisation, they are generally considered as high LET radiation. Neutrons interact with atomic nuclei from which they eject slow, densely ionising recoil particles which have short range and significant mass. The charged particles produced from neutron irradiation cause many ionisations as they traverse a cell and are very damaging to cells.

9.13 FRACTIONATION (1)

Problem:
Explain the following altered fractionation schemes and describe the radiobiological basis of each scheme:

a. Hyper-fractionation

b. Hypo-fractionation

c. Accelerated fractionation

Solution:

a. Hyper-fractionation refers to the use of smaller fraction sizes, multiple daily treatments, a higher total dose of radiation and a total treatment time that is about the same duration as for conventional radiotherapy. The aim is to further reduce late effects, while achieving the same or better tumour control with the same or slightly increased acute effects.

b. Hypo-fractionation uses a relatively high dose per fraction in a smaller number of fractions, given over a shorter period compared with conventional fractionation. This regimen may be more convenient for patients and less resource-intensive than the conventional fractionation schedule, with proven radiobiological effectiveness. However, this hypo-fractionation may result in higher late toxicities. This scheme is often used in advanced radiotherapy techniques such as SRS/SBRT. It is also used in palliative situations where one is attempting to rapidly relieve a particular symptom of a malignancy without having the patient go through a protracted course of radiotherapy when cure is not possible.

c. Accelerated fractionation refers to the use of a fraction about the same size (or slightly smaller) as in conventional fractionation, multiple daily treatments, a shorter overall treatment time and a total dose about the same (or slightly lesser) than the conventional radiation schema. The basic idea with this approach is to overcome the effects of tumour repopulation by shortening the overall time. Accelerated fractionation can increase the acute side effects and may require a break or reduced dose for patient tolerance. The late effects are not affected by accelerated fractionation because the total dose and fraction size are the same (or similar) with standard fractionation schema.

9.14 FRACTIONATION (2)

Problem:

a. Radical radiotherapy is given in small fractions over several weeks. Explain the rationale of this approach.

b. On the same graph, sketch the survival curves of cell lines irradiated with photons as:

  i. Single-fraction radiotherapy

  ii. Fractionated radiotherapy

c. A similar cell line system is then irradiated with carbon ions. Describe how the survival curve would appear with fractionated and single-fraction radiation. Why?

Solution:

a. Irradiated cells develop sub-lethal damage (SLD). When given time, the damage can be repaired. The repair of SLD is better in normal tissue compared to tumour tissue. A small differential is seen with every fraction, and over many fractions, the difference is more apparent.

b. See Figure 9.11.

c. Carbon ion is a type of heavy charged particle, which contributes to mainly α component of cell killing (rather than the β component). There is no or minimal SLD repair of the cells. Therefore, there is minimal or no difference in the survival curves between fractionated and single-fraction radiation.

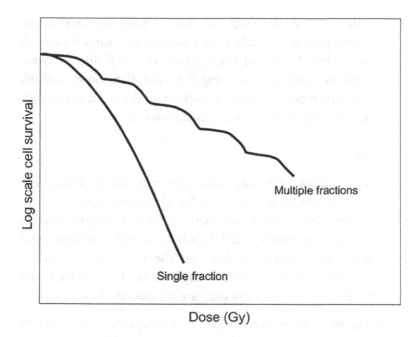

FIGURE 9.11 The survival curves of cell lines irradiated with photons as (i) single-fraction radiotherapy and (ii) fractionated radiotherapy.

9.15 TREATMENT INTERRUPTION

Problem:
It is unfortunately common for the schedule of radiotherapy treatments of patients to be interrupted and therefore to differ from that which was prescribed.

a. Explain the consequences of unplanned gaps on tumour cells during radiotherapy treatment.

b. Give the recommendations about how to compensate for unscheduled gaps in radiotherapy treatment.

c. A 40-year-old lady with colorectal cancer is being treated with radiotherapy. A dose of 45 Gy in 25 daily fractions over 5 weeks is planned. After ten fractions, treatment was cancelled for five fractions due to machine breakdown. It was planned to deliver the isoeffective tumour dose by increasing the size of the ten fractions to finish the treatment as scheduled. By using the 'equivalent dose' method, calculate required dose per fraction for the last ten fractions. Assume $\alpha/\beta = 10$ Gy for colorectal cancer.

Solution:

a. Repopulation of tumour cells may occur. The unplanned gap during radiotherapy will cause the cells within a tumour to proliferate. This can lead to an increase in the number of cells that the radiation needs to kill. In addition, there is evidence that repopulation can be accelerated in the treated tumours so that this effect can be significant. An increase of overall treatment time (due to the gap) will decrease the efficacy of radiotherapy.

b. The main recommendation is to compensate for a gap by continuing to deliver the same number of fractions in the same overall time, by inserting the missed fractions (using the original dose per fraction) either during weekends and/or as extra fraction on previously treatment days. When multiple fractions are given in a day, the interval between them should be at least 6 hours. If the break in treatment occurs early in the schedule, then two fractions per day can be used. However, if there are not enough days left on the original schedule to catch up using two fractions on each remaining days, three fractions per day can be considered subjected to the availability of resources.

c. Calculate the tumour equivalent total doses in 2 Gy fractions (EQD_2) from the planned remaining 15 fractions using

$$EQD_2 = D \cdot \frac{d + \left(\dfrac{\alpha}{\beta}\right)}{2\ \text{Gy} + \left(\dfrac{\alpha}{\beta}\right)}$$

where D = total dose and d = planned dose per fraction. We get

$$EQD_2 = 15 \times 1.8 \cdot \left(\frac{1.8 + 10}{2 + 10} \right)$$

$$= 26.5 \text{ Gy}$$

Now, estimate the dose per fraction (x) so that delivering ten fractions of this size gives an EQD_2 of 26.5 Gy to the tumour:

$$EQD_2 = EQD_2 = 10x \cdot \frac{d + \left(\dfrac{\alpha}{\beta} \right)}{2 + \left(\dfrac{\alpha}{\beta} \right)}$$

Substituting $EQD_2 = 26.5$ Gy and $\alpha/\beta = 10$ Gy into the above equation,

$$26.5 = 10x \cdot \frac{x + 10}{2 + 10}$$

Rearrange the above equation, we obtain

$$318 = 10x^2 + 100x$$

$$10x^2 + 100x - 318 = 0$$

Solving the quadratic equation,

We obtain $x = 2.5$

Thus, we would have to give fraction of 2.5 Gy for the last 10 fractions, a total of 25 Gy, in order to achieve the same tumour effect.

CHAPTER 10

Radiation Protection and Safety in Radiotherapy

10.1 DOSE QUANTITIES IN RADIATION PROTECTION

Problem:

Define the following dose quantities with their units used in radiation protection aspects:

a. Absorbed dose D_T

b. Equivalent dose H_T

c. Effective dose E

d. Deep dose equivalent H_p (10)

e. Shallow dose equivalent H_p (0.07)

DOI: 10.1201/9780429159466-10

177

Solution:

a. Absorbed dose D_T is a measure of the energy deposited in a medium by ionising radiation. It is equal to the energy deposited per unit mass of a medium, and so has the unit joules (J) per kilogram (kg), with the adopted name of Gray (Gy) where $1 \text{ Gy} = 1 \text{ J·kg}^{-1}$.

b. Equivalent dose H_T is a measure of the radiation dose to tissue where an attempt has been made to allow for the different relative biological effects of different types of ionising radiation. The equivalent dose in tissue T is given by the expression:

$$H_T = \sum W_R . D_{T,R}$$

where $D_{T,R}$ is the absorbed dose averaged over the tissue or organ T, due to radiation R. Equivalent dose is measured using the sievert (Sv).

c. Effective dose H_E is the sum of the weighted equivalent doses in all the tissues and organs of the body. It is given by the expression:

$$H_E = \sum W_T . H_T$$

where H_T is the equivalent dose in tissue or organ T, and W_T is the weighting factor for tissue T. Effective dose is measured using the sievert (Sv).

d. Deep dose equivalent H_p (10) is the dose equivalent at a tissue depth of 1 cm (1000 mg.cm^{-2}) due to external whole body exposure to ionising radiation. Deep dose equivalent is measured using the sievert (Sv).

e. Shallow dose equivalent H_p (0.07) is the external exposure dose equivalent to the skin or an extremity at a tissue depth of 0.07 mm (7 mg·cm^{-2}) averaged over an area of 1 cm^2. Shallow dose equivalent is measured using the sievert (Sv).

10.2 RADIATION SAFETY PRINCIPLES IN RADIATION ONCOLOGY

Problem:
Define the following principles which are used in radiation safety: justification, optimisation and the ALARA principle. Explain briefly how they are implemented within a radiation oncology department.

Solution:
Justification means that a clinical practice that involves exposure to ionising radiation should only be adopted if it provides sufficient benefit to the particular person. The benefit outweighs any detrimental effect that could occur to the person.

All sources of ionising radiation should have the best possible safety systems set in place so that magnitude and likelihood of exposure is kept as low as reasonably achievable (ALARA) taking into account the local economic and social factors being taken into account.

Optimisation means that the radiation dose is such that exposure to the ionising radiation provides the maximum benefit while minimising the risks. Within a radiation oncology environment, optimisation is achieved by exposure of the PTV to be consistent with the dose prescription by the radiation oncologist. In addition, the dose to the OAR and the healthy tissue is kept as low as possible. This is done within the treatment planning system. A part of this process includes the independent review of the radiation dose for the patient by checking DVH parameters with comparison to published dose limits such as those specified by Quantitative Analyses of Normal Tissue Effects in the Clinic (QUANTEC). Accuracy of these doses is also achieved by performed independent dose calculations and/or measurements of the treatment plan.

The optimisation of CT imaging doses is achieved in a number of ways included ensuring the FOV is optimum for the patient, accurate patient set-up to minimise the chance of needing to reimage and using the correct imaging protocol on the CT scanner.

10.3 METHODS OF RADIATION PROTECTION

Problem:
State and explain three basic methods used to reduce external radiation hazard.

Solution:
Time – Simply reducing the amount of time spent near or in contact with any source results in a proportionate reduction in dose.

Distance – Exposure decreases with distance according to the inverse square law, by which the radiation intensity varies inversely with the square of the distance from a source.

Shielding – Proper shielding can result in an exponential reduction of dose for gamma emitters and a near-total reduction for beta emitters.

10.4 PREGNANT PATIENT

Problem:
A 22-week pregnant patient has been diagnosed with breast cancer. It has been decided to commence radiation treatment to the breast. It is expected the foetus will remain outside the primary beam during the treatment.

a. What would be the sources of the radiation to the foetus, if the patient is treated with a 6 MV photon?

b. What steps should be taken to reduce the dose received by the foetus?

c. How could the radiation dose to the foetus be estimated?

Solution:

a. Photon leakage through treatment head, radiation scattered from the collimators and beam modifiers, radiation scattered within the patient from the treatment beams.

b. Modify the usual treatment technique, by changing field angles (avoid divergence), optimise field size, avoiding physical wedges that produce more scatter, choosing different radiation energy (avoid photons with $E > 10$ MV because of neutron production). Use suitable shielding device which usually involves the use of high Z materials. All aspects of shielding device must be carefully planned to eliminate the possibility of injury to patients or personnel. Leakage can be reduced by employing secondary shielding blocks. In linear accelerators with tertiary MLC, the MLC bank can be used for shielding by rotating the collimator appropriately. Finally, dose reduction starts with CT scanning for planning where the length of the scan should be reduced. If image guidance is used during treatment, electronic portal imaging during treatment delivery will carry no additional dose.

c. The radiation dose to the foetus can be estimated by performing dose measurement using suitable water or human phantom. Points of dose estimation should be selected that will reflect the range of dose throughout the foetus. Three points commonly used are the fundus, symphysis pubis and umbilicus.

10.5 CONTROLLED AND SUPERVISED AREA

Problem:
What are controlled and supervised areas? Give two examples each of such areas in a radiotherapy department.

Solution:
Controlled area is a work area where specific protection measures and safety provisions could be required for controlling normal exposures or preventing the spread of contamination during normal working condition and preventing or limiting the extent of potential exposures. Annual dose received by a worker in this area is likely to exceed 3/10 annual occupational dose limit.

The area must be demarcated with radiation warning signs and legible notices must be clearly posted. Examples of controlled area are a linac bunker and brachytherapy source preparation room.

Supervised area is a work area for which the occupational exposure conditions are kept under review even though specific protective measures and safety provision are not normally needed. The area must be demarcated with radiation warning signs and legible notices must be clearly posted. Examples are treatment console and areas surrounding treatment or simulation room.

10.6 EXTERNAL AND INTERNAL EXPOSURE

Problem:
Discuss the differences between external and internal exposures and the implications for radiation safety.

Solution:
External exposure: Radiation is reaching a person from outside through the skin. This is typically from a radiation source that can be turned off like an X-ray unit. The radiation from this source may cause damage in an organism while it is turned on. In general, no radioactive is left in the body. Hazards of this type may occur in radiology or radiotherapy departments.

Internal exposure: This occurs most commonly after the incorporation (e.g., breathing in, consuming with food and absorbing through the skin) of radioactive isotopes. The radioactivity remains in the organism until the isotope has decayed (physical half-life) or until it is excreted (e.g., in urine or during exhalation). These hazards may be present in laboratories or nuclear medicine departments.

10.7 RADIATION SOURCES FROM THE LINEAR ACCELERATOR

Problem:
List the four different types of radiation which may be present in a radiotherapy bunker containing a linear accelerator.

Solution:

- Primary radiation is the radiation that is directly emitted from the linear accelerator through the collimator/MLC.

- Scatter radiation is generated by scattering of the primary radiation beam as it interacts with the different materials such as the beam collimation, the patient, the treatment couch and the air.

- Leakage radiation is the radiation that escapes through the head of the gantry of the linear accelerator.

- Neutron radiation is created for photon beams when the energy is greater than 10 MV. These are mainly created in materials with high atomic number such as the target and collimation systems.

10.8 NEUTRON SHIELDING

Problem:

When medical accelerators are operated above 10 MV, neutrons will be produced.

a. What interaction causes the production of neutrons? Explain.

b. Explain the shielding requirement for the production of neutrons in a linear accelerator bunker.

Solution:

a. The photonuclear interaction (gamma, *n*).

b. Since practical therapy rooms need concrete walls at least two X-ray tenth value layers thick, adequate concrete shielding for the photons will always be adequate for the neutrons as well. However, great care should be taken if iron or lead is used for part of the shielding. Not only the attenuation of neutrons is poorer but interactions in iron or lead can create problem. Hydrogenous material provides best shielding for neutrons. In general, sandwich arrangements of lead, steel and polyethylene or concrete (hydrogenous material) are needed to provide adequate shielding for neutron.

The maze must be sufficiently long and appropriately designed to reduce neutron flux. A neutron door including paraffin wax (which is a good neutron moderator) and boron (which is a good absorber of thermal neutrons) may be required.

Neutrons can result in activation products. These are typically short-lived with half-life of the order of minutes. They mostly occur in the treatment head and closing the collimators prior to entering the room can reduce dose to staff. It is also advisable to let activation products decay prior to entering the room (>10 min) after prolonged use of high-energy photons (e.g., for commissioning).

10.9 SHIELDING BARRIER

Problem:

a. Radiation treatment facilities are comprised of primary and secondary barriers. Explain primary and secondary barriers.

b. Figure 10.1 shows a proposal of a bunker layout for a 6 MV linear accelerator facility. The area behind the wall will be used as a storeroom. The linear accelerator operates at a maximum dose output rate of 600 MU min^{-1}, use factor, U = 0.25. The expected workload is 35 patients per day (8 hours), 5 days per week and a dose of 2 Gy delivered at the isocentre per patient. Is the shielding adequate?

Solution:

a. Radiation treatment facilities are comprised of primary and secondary barriers. A primary barrier is required for the wall and roof where the main radiation beam reaches them.

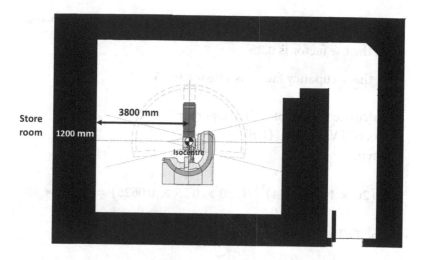

FIGURE 10.1   A proposal of a bunker layout for a 6 MV linear accelerator facility.

If the facility is located above any accessible area, the floor will need to be a primary barrier too. The remaining walls which are much thinner than the primary barriers are called secondary barriers. The secondary barriers will protect against scattered and leakage radiation.

b. To calculate the attenuation of the barrier, first calculate the attenuation of the barrier using:

$$B = \left(P(d + SAD)^2 \right) / WUT$$

where

B: the attenuation of the barrier

P: the allowed dose limit outside the barrier for a public area is 1 mSv per annum, (1/50) = 20 µSv week^{-1}

d: the distance from the isocentre to the point of interest on the far side of the barrier in meter is $(3800 + 1200)/1000 = 5$ m

SAD: source to axis distance in meter is 1 m

W: the workload in Gy week^{-1} at 1 m is $(35 \times 2 \times 5) = 350$ Gy week^{-1}

U: the use factor is 0.25

T: the occupancy factor is $1/16 = 0.0625$

To calculate the no. of TVLs required:
No. of TVLs = $log_{10}(1/B)$
Then,

$$B = \left(20 \times 10^{-6}(5 + 1)^2\right)/(350 \times 0.25 \times 0.0625) = 1.317 \times 10^{-4}$$

No. of TVLs = $log_{10}[(1/(1.317 \times 10^{-4})) = 3.88]$

The TVL for 6 MV X-ray in concrete (density 2350 kg m^{-3}) is 343 mm and the number of TVLs for the proposed wall is (1200 mm/343 mm) = 3.50. The proposed thickness of the wall is insufficient to shield the radiation and keep the dose below allowed dose limit. Therefore, the required thickness for primary barrier is (3.88 × 343 mm) 1331 mm.

10.10 RADIOTHERAPY BUNKER MATERIALS

Problem:
List the main materials that are used in radiotherapy bunker shielding construction and briefly provide comments on their typical use.

Solution:
The main materials used in radiotherapy bunker shielding constructions are as follows:

- Normal concrete: average density $\rho = 2.35$ g/cm^{-3}

- High-density concrete: average density $\rho = 5$ g/cm^{-3}

- Steel: average density $\rho = 7.9$ g/cm^{-3}

- Lead: average density $\rho = 11.3$ g/cm^{-3}

- Borated polyethylene $\rho = 0.94$ g/cm^{-3}

The most commonly used material for bunker design is concrete. However, in situations where space is limited, then the use of the higher density concrete, steel or lead may be used. Steel can be readily fixated on the current walls or as part of concrete. Lead while having the highest density can be difficult to have in the thicker flat sheets and it may be more expensive compared to the other materials. The use of the higher density materials would be needed if a bunker is being retrospectively upgraded to include a linear accelerator with higher energy X-ray beams, e.g., addition of 10 or 18 MV.

Borated polyethylene has a high neutron cross section and so will be used in the construction of neutrons doors for linacs with X-ray beams above 10 MV.

10.11 DOSE LIMITS

Problem:

a. What is the International Commission on Radiological Protection (ICRP) recommended effective dose limits for:

   i. Occupational dose?

   ii. Public dose?

b. What is the basis for a separate and reduced radiation dose limit for members of the general public compared to that for radiation workers?

Solution:

a.

   i. 20 mSv per year (averaged over 5 consecutive years), 50 mSv in a single year.

ii. 1 mSv per year (averaged over 5 consecutive years), 5 mSv in a single year.

b. The separate and considerably lower limit of 1 mSv per year for the general public takes into account the fact that this group includes children who may be more radiosensitive and who also have a longer time for possible detrimental effects to appear following irradiation.

10.12 RADIATION INCIDENT

Problem:
Assume the following scenario: A shielding block has been omitted for the eye of a patient treated with external beam therapy for 1 day. Discuss:

a. The consequences for the patient

b. The actions to be taken

c. Methods to prevent this accident from occurring again in the future

Solution:

a. Consequences for the patient:
The consequences for the patient will very much depend on the dose per fraction, and anatomical part of the eye that was being irradiation. ICRP Publication 118 states that the threshold dose for radiation-induced cataracts is relatively low, which is approximately 0.5 Gy for both acute and fractionated exposures. The omission of the block may increase the likelihood of cataract formation. Consideration must also be taken on the circumstances of the patient (age, profession, status of the other eye) and if other structures (e.g., the optic nerves) are also irradiated which may cause ocular complications.

b. Actions:

- The incident has to be documented, with dose estimates to the eye. A mock-up of the situation (e.g., dose measurement on an anthropomorphic phantom) may help to reconstruct doses.

- Inform radiation oncologist who should explain the incident to the patient. A thorough eye examination or vision test may be conducted by the clinician.

- In most countries there is a requirement for reporting incidents like this to the regulatory authority. The licence conditions for your operation will detail these requirements.

- It is also recommended to report incidences such as this to national and international data bases such as The Radiation Oncology Safety Information System (ROSIS) or IAEA Safety in Radiation Oncology (SAFRON). This will ensure others can learn from incidents.

c. Preventative measures:

- Encode treatments with blocks in the record and verify (R&V) system. Use MLC (instead of manual blocks) whenever possible as it is more automated and encoded in the R&V system.

- Highlight the use of blocks in treatment sheet.

10.13 DEVELOPMENT OF A RADIATION MANAGEMENT PLAN

Problem:
You have been given the responsibility of developing and setting up a radiation management programme (RMP) for a new radiotherapy centre consisting of three multimodality linear accelerators, CT scanner and a kilovoltage X-ray unit. List ten key steps you would use to develop and implement the RMP:

Solution:

- Ensure that the RMP and radiation safety measurements comply with local and national standards as set by regulatory bodies.

- Full radiation survey performed around all of the radiotherapy equipment to confirm dose levels as compared to the bunker design in both the controlled and uncontrolled regions.

- Personnel radiation monitors provided to all of the staff who are classified as radiation workers.

- A review of the dose levels measured for all personnel monitors with appropriate action levels determined when dose values are exceeded.

- Set-up of radiation safety committee with representatives from the each of the groups of radiation workers in the department.

- Radiation safety protocols, manuals and training provided to all staff.

- Incident reporting process within the hospital.

- Access to radiation monitoring equipment such as portable radiation survey meters.

- Dose calibrations are performed on the linear accelerators per national standards (based on IAEA TRS398 CoP or other accepted protocol).

- Quality assurance programme developed for all radiotherapy equipment with appropriate tolerances and action levels.

- Physical safety measures set up at the linear accelerators and CT scanner such as audio-visual system, a last-person-out-system, ionising radiation safety signs and warning lights.

- Safety checks performed at the time of the linac warm-up of these physical safety measures to ensure they are functioning correctly.

- Appointment of a radiation safety officer.

- Participation in dosimetry audits such as those offered by the IAEA.

- Commissioning records have been developed for the radiotherapy equipment including signatures of the responsible physicists.

- Posters located in the department for ensuring the safety of pregnant patients and staff members.

- An annual audit of all the radiation safety measures including spot checks of the radiation survey.

- Calibration records for your reference dosimetry equipment such as the reference ionisation chamber and electrometer which is sent to the SSDL or PSDL.

- All emergency off switches on the equipment work correctly.

- Documentation for the safe use of the radiotherapy equipment.

- Training records for all radiation workers in the use of the radiotherapy equipment.

- Record of significant radiation incidents and radiation accidents are reported to the local regulatory bodies.

Bibliography

Andreo, P., et al., *Absorbed Dose Determination in External Beam Radiotherapy, An International Code of Practice for Dosimetry Based on Standards of Absorbed Dose to Water, Technical Report Series No. 398.* 2000, Vienna: International Atomic Energy Agency.

Andreo, P., et al., *Fundamentals of ionizing radiation dosimetry.* 2017. New Jersey, USA: John Wiley & Sons.

Benedict, S.H., et al., Stereotactic body radiation therapy: The report of AAPM Task Group 101. *Medical Physics*, 2010. **37**(8): p. 4078–4101.

Bissonnette, J.P., et al., Quality assurance for image-guided radiation therapy utilizing CT-based technologies: A report of the AAPM TG-179. *Medical Physics*, 2012. **39**(4): p. 1946–1963.

Commissioning of Radiotherapy Treatment Planning Systems: Testing for Typical External Beam Treatment Techniques. 2008, Vienna: International Atomic Energy Agency.

Das, I.J., et al., Report of AAPM Task Group 155: Megavoltage Photon Beam Dosimetry In Small Fields and Non-Equilibrium Conditions. *Medical Physics*, 2021. **48**(10): p. e886–e921.

Design and Implementation of a Radiotherapy Programme: Clinical, Medical Physics, Radiation Protection and Safety Aspects. 1998, Vienna: International Atomic Energy Agency.

Devic, S., Radiochromic film dosimetry: Past, present, and future. *Physica Medica*, 2011. **27**(3): p. 122–134.

Dosimetry of Small Static Fields Used in External Beam Radiotherapy. 2017, Vienna: International Atomic Energy Agency.

Fraass, B., et al., American Association of Physicists in Medicine Radiation Therapy Committee Task Group 53: Quality assurance for clinical radiotherapy treatment planning. *Medical Physics*, 1998. **25**(10): p. 1773–1829.

Frenière, N., COMP report: CPQR technical quality control guidelines for brachytherapy remote afterloaders. *Journal of Applied Clinical Medical Physics*, 2018. **19**(2): p. 39–43.

Hanley, J., et al., AAPM Task Group 198 Report: An implementation guide for TG 142 quality assurance of medical accelerators. *Medical Physics*, 2021. **48**(10): p. e830–e885.

Hill, R., et al., Advances in kilovoltage X-ray beam dosimetry. *Physics in Medicine*, 2014. **59**(6): p. R183.

Hill, R., et al., Australasian recommendations for quality assurance in kilovoltage radiation therapy from the Kilovoltage Dosimetry Working Group of the Australasian College of Physical Scientists and Engineers in Medicine. *Australasian Physical & Engineering Sciences in Medicine*, 2018. **41**(4): p. 781–808.

ICRP, *The 2007 Recommendations of the International Commission on Radiological Protection*. 2007, Ontario, Canada: International Commission on Radiological Protection.

Introduction of Image Guided Radiotherapy into Clinical Practice. 2019, Vienna: International Atomic Energy Agency.

Joiner, M.C. and A.J. van der Kogel, *Basic Clinical Radiobiology*. 2018. Boca Raton, Florida: CRC Press.

Keall, P.J., et al., The management of respiratory motion in radiation oncology report of AAPM Task Group 76 a. *Medical physics*, 2006. **33**(10): p. 3874–3900.

Khan, F.M. and J.P. Gibbons, *Khan's the Physics of Radiation Therapy*. 2014. Philadelphia, USA: Lippincott Williams & Wilkins.

Klein, E.E., et al., Task Group 142 report: Quality assurance of medical accelerators a. *Medical Physics*, 2009. **36**(9Part1): p. 4197–4212.

Koken, P.W. and L.H. Murrer, Total body irradiation and total skin irradiation techniques in Belgium and the Netherlands: Current clinical practice. *Advances in Radiation Oncology*, 2021. **6**(4): p. 100664.

Ma, C.M., et al., AAPM protocol for 40–300 kV x-ray beam dosimetry in radiotherapy and radiobiology. *Medical physics*, 2001. **28**(6): p. 868–893.

Mayles, P., A. Nahum, and J.-C. Rosenwald, *Handbook of Radiotherapy Physics: Theory and Practice*. 2007. Boca Raton, Florida: CRC Press.

Metcalfe, P., T. Kron, and P. Hoban, *The Physics of Radiotherapy X-ray from Linear Accelerators*. 1997, Madison, WI: Medical Physics Publishing.

Mijnheer, B., et al., *Monitor Units Calculation for High Energy Photon Beams: Practical Examples*. 2001, Belgium: Estro Brussels.

Morales, J.E., et al., An experimental extrapolation technique using the Gafchromic EBT3 film for relative output factor measurements in small x-ray fields. *Medical Physics*, 2016. **43**(8Part1): p. 4687–4692.

Podgorsak, E.B., ed. *Radiation Oncology Physics: A Handbook for Teachers and Students*. 2005, Vienna: International Atomic Energy Agency.

Podgorsak, E.B., *Radiation Physics for Medical Physicists*. 2006, Berlin: Springer-Verlag.

Pomerleau-Dalcourt, N. and P. Basran, COMP Report: CPQR technical quality control guidelines for data management systems. *Journal of Applied Clinical Medical Physics*, 2018. **19**(5): p. 347–364.

Radiation Biology: A Handbook for Teachers and Students. 2010, Vienna: International Atomic Energy Agency.

Radiation Protection in the Design of Radiotherapy Facilities. 2006, Vienna: International Atomic Energy Agency.

Rivard, M.J., et al., Update of AAPM Task Group No. 43 Report: A revised AAPM protocol for brachytherapy dose calculations. *Medical Physics*, 2004. **31**(3): p. 633–674.

Rosenfeld, A.B., et al., Semiconductor dosimetry in modern external-beam radiation therapy. *Physics in Medicine & Biology*, 2020. **65**(16): p. 16TR01.

Solberg, T.D., et al., Quality and safety considerations in stereotactic radiosurgery and stereotactic body radiation therapy: Executive summary. *Practical Radiation Oncology*, 2012. **2**(1): p. 2–9.

Van Dyk, J., ed. *The Modern Technology of Radiation Oncology*. 1999, Madison, WI: Medical Physics Publishing.

Villarreal-Barajas, J.E., COMP report: CPQR technical quality control guidelines for treatment planning systems. *Journal of Applied Clinical Medical Physics*, 2018. **19**(2): p. 35–38.

Williams, J.R. and D.I. Thwaites, *Radiotherapy Physics in Practice*. 2000, Oxford, UK: Oxford University Press.

Printed in the United States
by Baker & Taylor Publisher Services